Managing
Your
Employees

T0206499

NATIONAL ASSOCIATION OF HOME BUILDERS
Business Management & Information Technology Committee
Single Family Small Volume Builders Committee

A Service of

NAHB

BuilderBooks™
National Association of Home Builders
1201 15th Street, NW
Washington, DC 20005-2800
(800) 223-2665
www.builderbooks.com

Human Resources Guide for Builders

Managing Your Employees

Theresa Minch	Executive Editor
Jill Tunick	Editor
Katy Tomasulo	Copyeditor
Jessica Poppe	Assistant Editor
E Design Communications	Cover Designer

BuilderBooks at the National Association of Home Builders

THERESA MINCH	Executive Editor
DORIS M. TENNYSON	Senior Acquisitions Editor
JESSICA POPPE	Assistant Editor
JENNY STEWART	Assistant Editor
BRENDA ANDERSON	Director of Fulfillment
GILL WALKER	Marketing Manager
JACQUELINE BARNES	Marketing Manager
GERALD HOWARD	NAHB Executive Vice President and CEO
MARK PURSELL	Executive Vice President Marketing & Sales
GREG FRENCH	Staff Vice President, Publications and Affinity Programs

ISBN 0-86718-581-3

Printed in the United States of America

Cataloging-in-Publication Data available at the Library of Congress

Disclaimer
This publication is designed to provide accurate and authoritative information in regard to the subject matter covered. It is sold with the understanding that the publisher is not engaged in rendering legal, accounting, or other professional service. If legal advice or other expert assistance is required, the services of a competent professional person should be sought.
   —From a Declaration of Principles jointly adopted by a Committee of the American Bar Association and a Committee of Publishers and Associations.

For further information, please contact:
BuilderBooks™
National Association of Home Builders
1201 15th Street, NW
Washington, DC 20005-2800
(800) 223-2665
Check us out online at: www.builderbooks.com

12/03 E Design/Circle/Bang 2000

Acknowledgments

We extend our sincere appreciation to the following reviewers and contributors who gave their valuable time and expertise to create an important resource for home builders:

Author of *NAHB Personnel Handbook for Small Volume Builders* (1996 edition):
Alan Clardy, Ph.D.

Contributor to *NAHB Personnel Handbook for Small Volume Builders* (1996 edition):
C.J. Sperl

Legal, Safety, and Personnel Advisors:
Callista Freedman
Dawn Harris
David S. Jaffe
Gary Komarow
Kate Kirkpatrick
Leon R. Rogers
Regina C.B. Solomon, CSP

1996 NAHB Single Family Small Volume Builder Committee Members:
James Christo
Jim Erwin
Norman Gerber
Jess Hall
Walter Rebmann
Dave Roberts
Tom Stephani
Chip Vaughn

1996 NAHB Business Management & Information Technology Committee Members:
Mike Carlton
Martin Freedland
Steve Maltzman
Steve McGee
Dottie Piazza
Alan Trellis
Bob Whitten
Rich Westlake

Builder Reviewers:
Dan Dierker
Kathy Smith
Gerrit Tuijtem

Reviewers/Content Contributors for 2004 Revision:
Cheri Boehmer
Callista Freedman
Lucy Katz
Leon R. Rogers
Michael Sivage
Pamela Smith
Bob Whitten

Table of Contents

SECTION ONE   Managing Human Resources in Your Business
Regulations, Programs, and Procedures

SECTION TWO   Sample Employee Handbook

SECTION THREE   Personnel Management Documents

Introduction

This manual is designed to help you manage the "people paperwork" of your business easily, consistently, and productively. It will help you establish, clearly communicate, consistently implement, and document effective human resources policies and procedures. A human resource management system helps you recruit and retain the caliber of employees that can make your business thrive. Strong human resource management benefits the employer and employees in the following ways:

- It helps prevent misunderstandings about job performance expectations. Misunderstandings can create employee ill-will, resentment, sabotage, and hostility, which in turn may lead to lower productivity and higher turnover.
- It encourages employee commitment to your business by showing that the employee will be treated fairly. By describing your general approach to managing employees, you indicate that there are rules and procedures your company follows (although you retain your discretion with disclaimers). When you do this, you create a sense of certainty and fairness in the workplace.
- It gives employees a sense of the value you offer as an employer. By describing the policies and benefits you offer, employees can see that working for you is worth more than just a paycheck.
- It makes it easier for you to manage. The policies you establish are, in a way, pre-made decisions. When an employee requests something or has a problem, the policies in the handbook provide guidance about what to do. You do not have to make a new decision every time. This saves you time and helps you treat your employees consistently.

Included in This Manual

This manual is divided into three sections:

SECTION ONE Managing Human Resources in Your Business
Regulations, Programs, and Procedures

This section addresses the legal and regulatory authorities that affect human resources policies and provides step-by-step systems for managing functions such as employment at-will, recruitment and hiring, employee benefits, compensation, and safety programs, among others.

SECTION TWO Sample Employee Handbook

You'll find a sample employee handbook in Section II and on the enclosed CD. Use this information to create a handbook that meets your company's specific needs.

SECTION THREE Personnel Management Documents

Section III and the enclosed CD contain a variety of general-purpose forms, checklists, sample agreements, and government forms you'll find useful for human resource management. Like the employee handbook in Section II, the documents (except for the government forms) can be adapted to suit your company's specific needs.

YOUR NEXT STEP

The information and materials included in this manual are written for the "typical" home building business; however, before using them, you should review the materials to see if they will work for your company. Since each business is unique, make sure you agree with the policies before using them. You may adapt these general guidelines to your specific requirements, but remember that your changes must meet federal and state laws and regulatory requirements.

CONSULT WITH YOUR LAWYER

The information in this manual is provided to help you manage your human resources legally and effectively. Given the fast-paced changes in labor law, it is possible that some items may be affected by modifications and revisions to relevant laws and regulations.

The information in this manual is not legal advice. Consult with competent legal counsel before adopting the policies and procedures in this manual and in the employee handbook. Because this information is provided as a general guideline and is not offered as legal guidance, neither the National Association of Home Builders nor the contributors can be held accountable for how you use this information, the changes or modifications you make to the suggested materials, or the results you achieve.

HOW TO USE THE CD IN THIS BOOK

The sample employee handbook text and the human resource management documents are provided on the enclosed CD. The handbook text and documents are provided so that you may modify and customize them for your business. *(**Note:** You cannot customize the government forms.)*

The handbook text is offered in Microsoft® Word 2000/Windows 98 (wd2000) and WordPerfect (wp) text files.

The CD will automatically start when you insert it into your CD-ROM drive. Click on the links to view the files in a browser. Or you can open the Microsoft Word or WordPerfect application on your computer and then open the files using the application. Save the files on your computer's hard drive and then modify them if you wish. (See "The Handbook on CD" on page 90 in Section II.) If you encounter technical problems, consult the user's manual for the application.

Managing Human Resources in Your Business

Regulations, Programs, and Procedures

SECTION ONE

Overview

This section addresses the basic rules, guidelines, and practices for managing the human resources functions of your business. Consider this information when setting up systems for hiring, payroll, benefits, employee relations and communications, safety, and related activities. This section explains the minimum administrative system you should install to efficiently manage your human resources.

WHAT ARE HUMAN RESOURCES FUNCTIONS?

Human resources management is the set of activities you use to recruit, retain, and motivate employees to perform successfully and help your business succeed. Human resources management has traditionally included the following areas:

- Organizational staffing and planning
- Compensation and benefits
- Recruitment and selection
- Training and orientation
- Safety
- Performance management and appraisal

Across all of these various functions, home builders must be concerned with:

- Staying in compliance with relevant federal, state, and local laws and regulations.
- Improving productivity and their employees' quality of work life.
- Avoiding and resolving confrontations with the workforce.

MINIMUM REQUIREMENTS

Employers must meet certain minimum requirements that govern the proper way to manage their human resources effectively and legally. While many of the federal regulations described in this section are requirements for operations with more than 15 employees, it is in the best interests of small-volume builders with fewer than 15 employees to know about and consider adopting these policies. Many state and local regulations are similar and/or identical to the federal regulations, but apply to employers with as few as five employees. You can better communicate company policies and prevent employee misunderstandings by knowing the laws that do not apply to you as well as those you are required to comply with. Further, many of the requirements that do not apply to you are sound business practices.

HOW TO USE SECTION ONE

1. Look through each topic to see whether you are covered by the applicable law(s) or regulation(s).
2. If so, study the material and make sure you are following the law or regulation. If necessary, correct or update your policies and procedures.
3. If you are not covered by the law, you may still want to consider implementing the policy discussed.

The decision planning worksheet and checklist on page 120 of Section III outlines the actions you may be required to take on each functional topic covered.

General Employment Policies

EMPLOYMENT AT WILL

Introduction

Historically, the basis for the typical employment relationship is the Employment At-Will doctrine. This doctrine holds that if a person is hired for an indefinite time and without an explicit contract, then that person is an at-will employee. This means that either party—the employer or the employee—can discontinue that employment relationship at any time for any reason, provided the reason does not violate any federal, state, or local statute, rule, or regulation, or breach an established public policy. This doctrine favors employers because it allows the employer to hire and fire without being second-guessed.

However, the At-Will doctrine has been challenged over the past several years. The adoption of various state employment laws and many state court rulings have given employees increasing opportunities to dispute unilateral employer actions, such as terminations, where such actions violate public policy (e.g., firing an employee for filing a workers' compensation claim).

Fair Treatment of Employees Is Essential

The uncontested authority of employers under the At-Will doctrine has gradually eroded due to a number of legal safeguards and protections for employees. As the policies and laws detailed in this manual demonstrate, it is imperative that employers treat all employees fairly and equitably in all aspects of human resources management.

When employers practice fair and legal employee treatment, at-will employment continues to provide them with certain specific rights regarding employment termination.

Action Steps

If you wish to preserve the at-will nature of employment with your employees, you must affirmatively declare that intention in all communications with your employees. There are three main domains in which you must establish this policy to preserve an at-will employment relationship:

1. Adopt the policy in your fundamental management and communication to your employees. For example, do not refer to any employee as a "permanent" employee.
2. Describe your at-will employment policy during new-employee orientation. (See "New Employee Orientation" later in this section.)
3. Include an at-will employment statement in any employee handbook you create.

Sample At-Will Notice

The sample employee handbook in Section II includes an employment at-will disclaimer. You may adopt the disclaimer as is or modify it with advice from a lawyer to meet your needs. The disclaimer protects your right to enforce an at-will policy by explaining that the employee handbook is not an employment contract, that employment is not for a fixed term or definite period of time, and that employment may be terminated by the employee or the employer at any time and for any reason.

Guides for At-Will Disclaimer Notices

There are several important rules for adopting and communicating an employment at-will disclaimer policy. The disclaimer should be:

- Specifically labeled as such, in clear and bold type.
- Located in a distinct and conspicuous position in the handbook, preferably on a page by itself at or near the front of the handbook.
- Written in terms that your employees can understand, but not in such a simplistic form that you lose the at-will protections.

In addition, the disclaimer should specify that:

- The handbook contains general policies that do not represent a contractual obligation.
- The policies and procedures in the handbook are general guidelines, and management reserves the right to administer them at its own discretion and judgment. That is, management can deviate from these policies at any time or change the policies without notice.
- Either the employer or the employee can terminate employment at any time, without notice. Further, even though grounds for dismissal are noted, those grounds are not all-inclusive and the employee can be terminated at any time for any legal reason.
- No representative of the company has the authority to enter into any agreement for employment for any specified period of time.

Note: *You should check your state's treatment of the at-will employment relationship to ensure that your employee handbook mentions the appropriate disclaimers for your state.*

EQUAL EMPLOYMENT OPPORTUNITY

Introduction

The basic purpose of equal employment laws is to create a working environment that is free from discrimination based on race, color, religion, sex, age, national origin, marital status, Vietnam Era Veteran's status, or physical or mental disability to the extent required by the law. Employers are prohibited from discriminating against their employees in the setting and management of employment terms and conditions, such as hiring, firing, compensation, training, and in all other areas of employment.

So, if it appears that you are treating members of a protected group (such as a minority group or women) differently from the way you generally treat all other employees in the terms and conditions of employment, you may be found to have violated equal employment laws.

Why Equal Employment Is Good Business

There are several reasons why equal employment opportunity is a policy that is good for business:

- By treating all employees fairly and consistently, you build employee commitment and loyalty to your business.

- By dealing with employees based on what they do—not who they are—you can develop a work-force of high talent and capability.
- An equal opportunity employer can benefit in the marketplace by avoiding negative publicity and lawsuits about poor employment practices that drive away business and good employees.

Legal or Regulatory Basis

The following federal laws create the legal basis for employers to adopt anti-discrimination policies, procedures, and practices:

- Title VII of the Civil Rights Act of 1964 (with amendments)
- Age Discrimination in Employment Act
- Americans With Disabilities Act of 1990
- Pregnancy Discrimination Act
- Rehabilitation Act
- Vietnam Era Veterans' Readjustment Assistance Act
- Executive Order 11246
- Uniform Guidelines
- Equal Pay Act

Are You Covered?

Basically, you are covered by the various federal Civil Rights Act laws if you employee 15 or more employees. However, you should note that state and local civil rights laws frequently cover employers with less than 15 employees.

If you are not covered by these employment laws, it is still good business practice to manage and direct your employees as if you were covered. Many states allow employees to bring state tort actions called "wrongful discharge" or "constructive discharge" against their employers with less than 15 employees for discriminatory actions prohibited by the Equal Employment Opportunity laws.

Action Steps

An employer should develop policies and programs that demonstrate the employer's commitment to equal employment opportunity and that identify the steps and procedures the employer takes to avoid illegal discrimination.

In practical terms, these policies apply throughout the employment relationship, from recruitment, selection, promotion, and discipline to termination. Thus, the following actions are all recommended as appropriate and prudent steps to take in establishing your human resources management program.

1. Adopt a policy in favor of equal employment opportunity. A sample policy on equal employment opportunity is included in Section II. You should review that policy and formally adopt it for your organization. Depending on the structure of your business, this may involve having your board of directors approve the policy. That policy should be communicated to your managers, employees, job applicants, and others.
2. Use non-discriminatory recruiting ads. When you post notices of open positions, do not use terms, phrases, or language that indicate a preference for people of a certain gender, age, or racial or ethnic background. While you are not required to do so under the Equal Employment Opportunity laws, it's appropriate to note that you are an EEO employer.
3. Use job-related procedures for selecting new employees. To hire the most qualified person for the job, regardless of racial or ethnic background, age, or gender, you should use job-related selection

procedures. The same selection process should be used with all applicants who pass the initial candidate screening. All documentation should be signed and dated.

For example, use the same interview questions, test, or skills demonstration for each and every applicant for the same job.

4. Maintain good records. It is always a good idea to keep written documentation of any action taken involving an employee, be it a promotion or a termination. The record can be as simple as a few sentences, but the intent is to describe the job-related, non-discriminatory reasons behind the action.

Whenever you document an action, be sure to factually describe the action taken. Leave all opinions, personal impressions, and personal thoughts out of the document.

5. Notify employees of your job expectations. By indicating to employees in advance what you expect of them at work, you also signal what employee actions may or may not be acceptable. This indicates that you will treat employees based on what they do on the job, not who they are.

6. Treat everyone in the same manner. Treat people consistently. Nepotism and favoritism breeds employee discontent. A disgruntled employee may use inconsistent treatment of employees in the workplace as proof that an employer acted in a discriminatory manner.

7. Audit existing practices. Take a proactive role by frequently reviewing your existing employment practices. Employment laws and their interpretation can change. Are there things you are doing to discriminate, even unintentionally, against employees because of their race, sex, age, religion, or disability? Are your policies producing results that were fine years ago, but now, due to changes in employment laws, indicate that your policies are discriminatory?

State and/or Local Laws

In addition to the federal laws noted above, there also may be state or local laws in your area that create other "protected categories" of employees. Some possible additional categories include:

- Marital status
- Political preference
- Sexual orientation

Check with your state and local governing authority for more details.

If there are other protections, this means that you cannot legally make employment decisions based on these protected classifications.

AFFIRMATIVE ACTION

Introduction

In addition to the equal employment opportunity laws, you may also be required to prepare and file an affirmative action plan. This plan describes how you will make efforts to recruit, hire, and advance specific groups, such as women, persons with disabilities, or veterans.

Legal or Regulatory Basis

Executive Order 11246 is the main authority. In addition, the Rehabilitation Act of 1973 and the Vietnam Era Veterans' Readjustment Assistance Act also apply.

What Is Required

Executive Order 11246 prohibits businesses that work for the federal government (that is, are contractors) from engaging in discrimination on the basis of race, sex, religion, or national origin and requires them to take affirmative action in hiring and advancing minority group members and women.

The Rehabilitation Act requires contractors to take affirmative action to employ and promote individuals with disabilities.

The Vietnam Era Veterans' Readjustment Assistance Act requires contractors to take affirmative action in hiring and promoting Vietnam-era veterans and veterans who served on active duty during a war or campaign for which a campaign badge has been authorized.

Are You Covered?

Executive Order 11246 applies to any federal contractor with a federal contract, and federally assisted contracts, that exceeds an aggregate total of $10,000 in any 12-month period. When the $10,000 threshold is met, all contracts will be subject to the Order regardless of the amount of each individual contract.

The Rehabilitation Act applies to government contracts exceeding $10,000 that relate to personal property and non-personal services, including construction. This act applies to prime contractors and subcontractors.

The Vietnam Era Veterans Readjustment Assistance Act applies to government contracts of at least $25,000 that relate to personal property and non-personal services, including contractors. This act applies to prime contractors and subcontractors.

Written Affirmative Action Plans

Although each of these laws requires that contractors take affirmative action, a written affirmative action plan need only be drafted if you are a contractor with 50 or more employees *and* have government contracts of at least $50,000, or have government bills of lading that total or will total at least $50,000 in a 12-month period.

Note: *Blanket purchase agreements exceeding $50,000 annually constitute a single contract for the purpose of determining whether or not you are required to submit a written affirmative action plan.*

In addition, under the Vietnam Era Veterans' Readjustment Assistance Act, every federal contractor that has a contract of $50,000 or more *and* has 50 or more employees must include an equal opportunity clause in its contract.

An affirmative action plan must be written within 120 days of the award of the contract, and, in general, must include the following:

- Company's equal opportunity policy statement, including:

  - ▲ The chief executive officer's attitude on the subject matter.
  - ▲ That the contractor will recruit, hire, train, and promote people in all job titles.
  - ▲ That the contractor will ensure that all other human resources actions are administered without regard to race, color, religion, sex, age, national origin, Vietnam Era Veteran's status, or physical or mental disability.
  - ▲ That all employment decisions are based only on valid job requirements.
  - ▲ That employees and applicants shall not be subjected to harassment, intimidation, threats, coercion, or discrimination because they have engaged in or may engage in filing a complaint, assisting or participating in an investigation, hearing, or other activity related to the administration of the acts; opposing any act or practice unlawful under the acts; and exercising any other right protected under the acts.

- Identification and analysis of the current work force, including any deficiencies inherent in minority employment, including:

  - ▲ Listing every job title ranked highest- to lowest-paid within each department.
  - ▲ Total number of incumbents for every job.
  - ▲ Listing male and female numbers per group of blacks, Spanish-surnamed Americans, American Indians, and Asians.
  - ▲ Wage rate and salary range for each position.

- Evaluation of the opportunity for hiring minority group personnel for your business, including:

  - ▲ Analysis of minority groups in all job categories.
  - ▲ Analysis of hiring practices during the past year, including recruitment sources, testing, and determining whether equal employment opportunity is being afforded in all job categories.
  - ▲ Analysis of upgrades, transfers, and promotions for each position during the past year to determine whether or not equal employment opportunity is being provided.
  - ▲ Identifying areas in which the contractor is deficient in hiring minorities and women.

- Provide specific detailed steps to guarantee equal employment opportunity keyed to the problems and needs of members of minority groups, including identifying areas where there are deficiencies and the development of specific goals and schedules for the prompt achievement of full and equal employment opportunities at all levels and in all jobs within the company.
- A table of job classifications, including job titles, principal and auxiliary duties, pay rates, and other applicable rates (if more than one rate is paid).
- Signature of the contractor's executive official.

SEXUAL HARASSMENT
Introduction

As part of the civil rights laws, it is illegal for employees to be sexually harassed as part of their employment with you. You should take precautions to guard against having harassment occur and to deal forcefully with any evidence of harassment.

What Is Sexual Harassment?

According to the law and regulations, sexual harassment can occur whenever there is behavior of a sexual nature that is unwelcome. This behavior becomes harassing by:

- Making continued employment and other human resources actions (such as raises or demotions) dependent on the other party either accepting or rejecting those advances.
- Creating an intimidating, hostile, or offensive working environment that unreasonably interferes with the employee's job performance or the work environment.

Sexual harassment of your employees may occur from managers who work for you, other employees, or others (such as subcontractors or suppliers) at the jobsite. You should act to prevent and correct such harassment immediately, regardless of the source.

Action Steps

Do the following to protect your business from sexual harassment problems:

1. Adopt and communicate a written policy against sexual harassment in your workplaces and stick to it. The policy should guarantee no retaliation for good-faith reporting of sexually harassing experiences. A sexual harassment policy is included in the sample employee handbook in Section II.
2. Specify an internal reporting procedure whereby an employee who believes he or she is being sexually harassed is encouraged to and may report that condition. As part of this procedure, the employee should be able to go to someone other than his or her boss.
3. Train your managers to identify and avoid sexual harassment problems.
4. When there is a complaint, investigate it thoroughly. The obligation to investigate generally is triggered by a supervisor's observation of inappropriate banter, conversations, or conduct, as well as by employee complaints.
5. Take appropriate action, up to and including terminating the harasser's employment if it is determined that harassment or other inappropriate conduct took place.
6. Where applicable, restore lost employment benefits and/or opportunities to the victim.
7. Prevent the misconduct from occurring again.

Components of the Investigative Process

Begin and conclude the investigation promptly. The length of time required depends on the complexity of the circumstances.

Choose an appropriate investigator with the knowledge that he or she may be called to testify at a trial. The investigator should have the ability to understand the various equal employment laws, should be objective and impartial, and should maintain confidentiality of the proceedings.

Document your investigative procedure as well as your factual findings. Avoid inserting any opinions, assumptions, or personal interpretations into the documentation as to whether or not harassment occurred.

Take corrective action. This includes immediately halting ongoing harassment, taking disciplinary action that is reasonably calculated to end the harassment, counseling affected employees, and training all employees about harassment and its consequences.

Note: *There is a recent trend among alleged sexual harassers to sue their employers for defamation and/or wrongful termination. Sloppy, unfair, and incomplete investigations add fuel to the fire for these kinds of suits.*

State and/or Local Laws

Some states, such as California, have specific and more detailed rules for employers to follow in dealing with sexual harassment. Consult with the appropriate authority for details specific to your area.

FAMILY AND MEDICAL LEAVE

Introduction

This policy requires that certain employers grant eligible employees unpaid leave for up to 12 weeks to care for certain serious medical conditions or newborn babies in their families.

Legal or Regulatory Basis

The Family and Medical Leave Act (FMLA) of 1993.

Are You Covered?

This law covers employers with 50 or more employees in the current or preceding calendar year.

Which Employees Are Covered?

To be eligible for this leave, an employee must have been employed by you for at least 12 months prior to the requested leave. In addition, the employee must have worked at least 1,250 hours during that 12-month period.

Calculation of 1,250 Hours

Use the principles of the Fair Labor Standards Act to calculate whether or not an individual has worked 1,250 hours. An employer must include all hours worked, including overtime hours and other duty time hours. Time paid but not worked (e.g., vacation and personal leave) is not counted. If an employer does not have actual records of hours worked, then an employee is presumed to have met the 1,250-hour eligibility requirement if he or she worked for the employer for at least 12 months.

How Is the 12-Month Period Determined?

You must choose one of the following methods to determine the 12-month period in which to calculate the amount of leave taken by your employees:

- Calendar year
- Any fixed 12-month period
- Rolling 12-month period from the date when an employee's first FMLA leave began

You may use any of the above methods, but it must be applied consistently and uniformly to all employees. If you decide to change the method used to calculate the 12-month period, you must give employees 60 days notice of the change and the transition must be done to ensure employees receive the greatest leave between the two methods.

What Notices Is an Employee Required to Give the Employer?

An employee must notify the employer of the need for FMLA leave. The employee must sufficiently explain the reason so that the employer can determine whether or not the leave qualifies under the Act.

An employee need not expressly assert rights under the FMLA or mention the FMLA to meet the notice requirements under the Act. It is sufficient for the employee to notify the employer of the need for such leave.

What Notices Is an Employer Required to Give Employees?

An employer must notify an employee within two business days of receipt of notice of need for FMLA leave that the leave will be counted toward FMLA leave (if it qualifies under the Act).

Notice to the employee may be oral or written. If oral, the employer must confirm the decision in writing no later than the day following payday. Sample Family and Medical Leave response forms appear on pages 123–130 in Section III.

What Are the Requirements Under the FMLA?

If you are covered by this law, the following benefit provisions apply:

- You are required to grant unpaid leave to an employee for up to 12 weeks in any 12-month period if any of the following conditions occur to that employee:

  - ▲ A newborn child enters the family.
  - ▲ A child is placed with the family due to adoption or foster care.
  - ▲ The employee assumes care for a child, spouse, parent, or dependent son or daughter over the age of 18 who has a serious medical condition.
  - ▲ The employee has a serious health condition that makes the employee unable to perform functions of the employee's job.

  A leave period includes intermittent leave periods and times with a reduced schedule.

  Once the employee returns to work, he or she must be given the same or an equivalent job of status and pay without loss of any compensation or benefits accrued prior to the leave—unless the employee is a key employee.

  An employer must notify an employee in writing of his or her status as a key employee and that there is a possibility of denying reinstatement. Notice must be given when the leave is requested or when leave begins.

  Note: *A key employee is salaried (paid on a salary basis rather than hourly), and is one of the 10 percent highest paid employees within 75 miles of the worksite. An employer may deny reinstatement only if the reinstatement would cause the company substantial and grievous economic injury.*

- You must notify the employee in writing whether or not he or she is eligible for family and medical leave. This must be done within two days after the employee requests leave or within two business days after the employer acquires knowledge that leave is being taken for FMLA qualification purposes, or the employee will be automatically deemed eligible.

  If the employer fails to notify the employee regarding leave designated as FMLA, it is not counted as FMLA leave.

- If you provide health care coverage and the employee is on the plan, you must maintain the employee's coverage while he or she is on leave on the same conditions as coverage would have been provided had the employee been continuously employed during the entire leave period. The employee is required to pay the health care premium he or she would have been paid if continuously employed on a timely basis (within 30 days of being due) or coverage can stop. However, coverage would need to be continued immediately upon the employee's return. If the employee does not return to work, the employer can act to recover any premium amounts still owed by the employee.

- You are required to communicate to employees their rights under this law in a manner they clearly understand. This communication must be posted in a conspicuous place and included in the employee handbook or some other written announcement of benefits.

- You may not discriminate against an employee using FMLA leave in any terms of employment, including incentive programs such as safety, attendance, or merit.

- You must make accommodations to employees who must receive intermittent care during the course of regular business hours. The employee is expected to minimize any disruptions to the employer's work schedule.

- You may require that paid leave (vacation, sick, personal leave) must first be used during the FMLA leave period and then the remaining FMLA leave time is unpaid.

What Is a Serious Health Condition?

A serious health condition entitling an employee to FMLA leave is an illness, injury, impairment, or physical or mental condition that makes the employee unable to perform the essential functions of his/her position. Such ailments as the common cold, influenza, earaches, upset stomach, minor ulcers, headaches (other than migraines), routine dental and orthodontia treatments, or periodontal disease are not included in this classification. A serious health condition does involve:

Inpatient care—Treatment in a hospital, hospice, or residential medical care facility, OR any subsequent treatment in connection with such inpatient care; or

- **Absence plus treatment**—Treatment by a health care provider that includes any one of the following:

  - Period of incapacity of more than three consecutive calendar days, and any subsequent treatment or period of incapacity relating to the same condition that also involves:

    - Treatment two or more times by a health care provider, by a nurse or physician's assistant under direct supervision of a health care provider, or by a provider of health care services (e.g., physical therapist) under orders of, or on referral by, a health care provider; or
    - Treatment by a health care provider on at least one occasion that results in a regimen of continuing treatment under the supervision of the health care provider.

  - Absences that are attributable to incapacity qualify for FMLA leave even though the employee or the immediate family member does not receive treatment from a health care provider during the absence, even if the absence does not last more than three days (e.g., employee unable to report to work due to onset of asthma attack; health care provider advised employee to stay home when the pollen count exceeds a certain level).

- **Pregnancy**—Any period of incapacity due to pregnancy or for prenatal care.
- **Chronic conditions requiring treatments**—A chronic condition is one which:

  - Requires periodic visits for treatment by a health care provider or by a nurse or physician's assistance under direct supervision of a health care provider.
  - Continues over an extended period of time (including recurring episodes of a single underlying condition).
  - May cause episodic rather than a continuing period of incapacity (e.g., asthma, diabetes, epilepsy).

- **Permanent/long-term conditions requiring supervision**—A period of incapacity that is permanent or long-term due to a condition for which treatment may not be effective. The employee or family member must be under the continuing supervision of, but need not be receiving active treatment by, a health care provider (e.g., Alzheimer's, a severe stroke, or the terminal stages of a disease).
- **Multiple treatments (non-chronic conditions)**—Any period of absence to receive multiple treatments (including any period of recovery there from) by a provider of health care services under orders of, or on referral by, a health care provider, either for restorative surgery after an accident or other injury, or for a condition that would likely result in a period of incapacity of more than three consecutive calendar days in the absence of medical intervention or treatment, such as cancer (chemotherapy, radiation, etc.), severe arthritis (physical therapy), or kidney disease (dialysis).
- **Substance abuse treatment**—An employer must grant leave for substance abuse treatment if conditions regarding serious health conditions are met. However, an employer may take an adverse employment action pursuant to an established policy if it is applied in a consistent and

non-discriminatory manner; the policy was previously communicated to employees; and the policy provides for disciplinary action and/or termination for substance abuse

Only if these conditions are met may an employee be terminated despite FMLA provisions.

What Is Treatment?

Treatment includes:

- Examinations to determine if a serious health condition exists and evaluations of those conditions.
- A regimen of continuing treatment. For example, a course of prescribed medication (e.g. antibiotics) or therapy requiring special equipment to resolve or alleviate the health condition.
- Restorative dental or plastic surgery after an injury or removal of cancerous growths if all other conditions of the regulations are met.

Treatment does *not* include:

- Routine physical examinations, eye treatments, dental examinations, acne treatments, or plastic surgery.
- Taking of over-the-counter medications such as aspirin, antihistamines, or salves; bed rest, drinking fluids, exercise; and other similar activities that can be initiated without a visit to a health care provider.

What Is a Health Care Provider?

Under the FMLA, a health care provider is defined as:

- A doctor of medicine who is authorized to practice medicine or surgery by the state in which the doctor practices.
- Podiatrists, dentists, clinical psychologists, optometrists, and chiropractors (limited to treatment consisting of manual manipulation of the spine to correct a subluxation that has been detected by x-ray) authorized to practice in the state and performing within the scope of their practice as defined under state law.
- Nurse practitioners, nurse-midwives, and clinical social workers who are authorized to practice under state law and who are performing within the scope of their practice.
- Christian Science practitioners listed with the First Church of Christ Scientist in Boston, Mass. An employee receiving treatment from this type of practitioner may not object to any requirement from an employer that the employee or family member submit to examination to obtain a second or third certification from a health care provider other than a Christian Science practitioner except as otherwise provided under state or local law or a collective bargaining agreement.
- Any health care provider from whom an employer or the employer's group health plan's benefits manager will accept certification of the existence of a serious health condition to substantiate a claim for benefits.
- Any health care provider listed above who practices in another country and is authorized to practice in that country.

Action Steps

If you are covered by the FMLA, you should take the following steps:

1. Considering the complexity of this act, you may wish to have your policy and procedures reviewed by a competent legal or management consultant.

2. Make sure you are providing all required communications to employees on a timely basis.
3. Create an administrative procedure to reply to all employee requests for family leave within two days. Standard response forms for these requests appear on pages 123–130 of Section III.
4. The law requires posting of FMLA rights even if an employer has no covered employees.
5. Train your managers and supervisors in the administration and application of FMLA leave policies and procedures. You may want to assign one individual to be responsible for administering the FMLA policies and procedures; this will help ensure that all appropriate notices are distributed.

State Issues

Your state may have its own regulations regarding family and medical leave. Some states require more stringent procedures and leave requirements than the FMLA does.

PREGNANCY
Introduction

Pregnancy should be treated in the same way your company treats any other temporary medical disability. Discrimination against women based on pregnancy, childbirth, or related conditions is prohibited. Pregnant women are entitled to the same leaves of absence, benefits continuation, and reinstatement processes as are other non-pregnant employees with medical disabilities.

Legal or Regulatory Basis

The Pregnancy Discrimination Act of 1978. The Family and Medical Leave Act (FMLA) may also apply.

Are You Covered?

The Pregnancy Discrimination Act covers all employers with 15 or more employees. The provisions of the FMLA only apply to employers with 50 or more employees.

Action Steps

1. Do not base any human resources decisions, such as hiring, promoting, or firing a woman, on the fact that she is pregnant.
2. Review your hiring, disciplinary, benefits, and leave policies to make sure pregnancy is treated no differently than any other disability or illness. For example, do not have a rule that prohibits women from returning to work for some preset time after childbirth. Also, do not force a woman to take leave if she and/or her doctor say she is able to work.
3. Carefully consider how to deal with pregnant women working in hazardous conditions. Simply removing a woman from a hazardous condition for what you believe to be a safety precaution for her or her unborn child may be a form of illegal employment discrimination based on sex if this removal limits or restricts the woman's employment opportunities. On the other hand, some states have laws for protecting pregnant women from hazards.
4. Check with the appropriate agency in your state to see if there are any specific laws governing pregnancy in the workplace.
5. Notify pregnant women of any specific risks or hazards they or their unborn children may be facing on the job.
6. Provide protective gear if the hazard cannot otherwise be eliminated or reduced.
7. Offer temporary assignments to other duties, with the understanding that the woman's employment opportunities will not be adversely impacted and that she will be returned to her normal position.

Related Matters

Under the FMLA, the condition of pregnancy and all that goes with it, such as morning sickness, are considered as serious health conditions. As such, you may be required to allow the employee to use family leave for pregnancy-related illness and disabilities.

Employers cannot discriminate against women who have had abortions.

EMPLOYING PERSONS WITH DISABILITIES

Introduction

In general, you may not discriminate in employment against a qualified person with a disability who can perform the essential functions of the job with or without reasonable accommodation. However, you are not required to hire a person who cannot perform the essential functions of a job even with reasonable accommodations.

Legal or Regulatory Basis

The Americans with Disability Act (ADA) of 1990.

Are You Covered?

The ADA covers employers with 15 or more employees.

What Is a Disability?

A "person with a disability" is defined as an individual who:

- Has a physical or mental impairment that substantially limits one or more major life activities, such as seeing or walking, or
- Has a record of such impairment (e.g., a history of heart problems), or
- Is regarded as having an impairment (even if the person does not have an impairment).

Temporary health problems, like a broken arm, appendicitis, and flu, are not substantial limitations and do not by themselves qualify the person as disabled. However, temporary conditions may turn into hidden disabilities. For example, an employee recovers from a broken ankle but has a permanent limp that substantially limits his or her ability to walk; that employee has a disability covered by the ADA. Some conditions such as current drug use, sexual behavior disorders, or compulsive behaviors such as gambling are not considered disabilities.

Action Steps

You should take the following actions to meet the requirements of this law:

1. Identify the "essential functions" of each position in your company. Essential functions are the ones you cannot operate without. (See the information about job descriptions under "Classifying Employees" later in this section for more about this topic.)
2. Use sound, job-related selection criteria for considering applicants for open positions. This means assessing whether applicants have the necessary qualifications for performing the essential functions of the job.

    Along the same lines, you may not ask about the nature or extent of a person's disability during the selection process. However, you may ask the applicant how he or she would perform the functions of the job, or if the employee can perform a particular function with or without a reasonable accommodation.

3. Make sure your screening procedures are adapted to the needs of applicants with disabilities. For example, you may need to write out interview questions for a person with impaired hearing.

4. For a qualified applicant with a disability, consider whether you can offer a reasonable accommodation that would allow the applicant to perform the essential functions of the job. For example, reasonable accommodations might include:

- Slight changes in work schedule or allowing a flexible starting and quitting time.
- Making facilities more accessible.
- Reassignment of some non-essential job duties.
- Acquisition of special equipment or tools.

An employer is only obligated to make accommodations that are reasonable; that is, not unduly harsh or expensive. The reasonableness of the accommodations depends on the size of the employer and the cost of the accommodations. An employer is not required by the ADA to create a new position as an accommodation.

5. You cannot require applicants to undergo medical examinations prior to making a job offer; however, you may make employment contingent on successfully completing a medical examination after a preliminary job offer has been extended. If you require a medical examination for one applicant, all applicants for the same or similar positions must also have medical examinations. Only use a medical or physical exam if the medical condition is important for job performance. The requirement for a medical or physical exam must be applied consistently across each job level. Tests for the illegal use of drugs are not considered to be a medical examination under the ADA and are not prohibited.

6. Keep medical records confidential; this can easily be done by keeping any such records in files separate from regular employee files and in a locked cabinet.

7. Use your employment handbook. Know and understand policies and procedures and apply them consistently to all of your employees.

What to do After an Employee With a Disability Is on the Job

If the employee requests a reasonable accommodation, document all requests, including:

- When and how the request was made.
- What steps were taken to assess different types of accommodations.
- Accommodations suggested by the employee and the employer.
- The reason a specific accommodation was rejected or chosen.

Your tasks as an employer are:

- Ask the individual requesting an accommodation to suggest an accommodation.
- Request medical documentation of the existence of a disability that requires an accommodation.
- Determine if the job may be modified.
- Discuss alternatives with the employee.
- Ask the employee which accommodation is desired.
- Give an employee's request appropriate consideration.
- Where an employee's suggestion cannot be accommodated, provide a reasonable alternative.
- Document what was discussed, what was suggested, what was offered (all parts of the offer), and whether the employee accepted or rejected the accommodation.

You do not need to create a new position.

INTERRELATIONSHIP AMONG THE ADA, FMLA, AND WORKERS' COMPENSATION LAWS

Introduction

The purposes of these three laws differ dramatically in effect on implementation:

- The ADA's purpose is to protect people from discrimination on the basis of disability.
- The FMLA serves to protect an individual's job while out on medical leave.
- The Workers' Compensation Law compensates workers for injuries that occur on the job.

Despite these diverse purposes, the laws interact and overlap in a manner which, when complying solely with one law, an employer may be exposed to potential financial risks of lawsuits by injured, ill, or disabled workers. An employer, therefore, must comply with whichever statute provides the greater rights to employees in that particular situation.

Interrelationships

It is likely that many, but clearly not all, employees eligible for leave under the FMLA will qualify as "disabled" under the ADA. For example, heart conditions requiring ongoing treatment, infection with the HIV virus, and most cancers are considered "serious health conditions" under the FMLA as well as "disabilities" under the ADA. Likewise, employees who sustain injuries covered by workers' compensation might have a "serious health condition" covered by the FMLA as well as a "disability" under the ADA. Thus, in dealing with employees who have any type of health-related problem, employers must analyze each employee's health-related problem under the FMLA and the ADA. Likewise, many employees who have health-related problems will suffer a reduction in hours of employment or experience a termination of their employment, entitling them to benefits under COBRA. (See "Continuation of Medical Insurance Coverage" later in this section.) For employers to fulfill their obligations under the FMLA, ADA, and COBRA, it is essential that each situation involving an employee who has a health-related problem, including problems covered by workers' compensation laws, be analyzed under each law to determine the exact obligations owed the employee.

As each injured, ill, or disabled worker case arises, it should be carefully scrutinized to determine which act applies to that set of circumstances. Depending on the level of injury, a worker injured on the job may fall under one or all three acts. Work-related injuries do not always cause physical or mental impairments severe enough to "substantially limit" a major life activity. It is, therefore, the nature of a worker's disability that will determine the level of rights the worker is entitled to under these laws.

For example, if the injury is of short duration (e.g., a sprained ankle), the injury meets the criteria for a workers' compensation claim, but does not meet the definition of disability under the ADA. But if that same injury created long-term residual or chronic disabilities, such as a limp that substantially limits the ability to walk, it does fall under the definition of a disability. Under the ADA, the employer is expected to make reasonable accommodation for the employee's return to work. However, when the employee is ready to return to work but is unable to perform his/her same position, the employee may be entitled to take medical leave under the FMLA in lieu of an alternative assignment (provided the employee has worked 1,250 hours over the prior 12-month period and satisfies the serious health condition requirement). But if an employee has already missed six months of work because of an illness or work-related injury, he or she may not qualify for FMLA leave.

An employer must be extremely careful when making employment decisions that adversely affect employees exercising their rights under the FMLA, since the decision could subject the employer to liability under the FMLA, and possibly under the ADA. For example, the ADA provides that an employer may not discriminate against an employee because of his or her disability and requires the employer to provide a "reasonable accommodation" to the employee's disability unless it can be shown that the accommodation would impose an undue hardship on the employer. The reasonable accommodation required by the ADA might include letting the employee work a modified work schedule or allowing the employee a flexible leave schedule. If the employee is not covered by the FMLA because he or she was not employed for a year or did not work 1,250 hours during the previous year, the employer would not violate the FMLA by denying the employee a medical leave. However, the employer might violate the ADA by denying a medical leave if medical leave was a part of the reasonable accommodations necessary to accommodate a physical or medical disability under the ADA.

Another area that merits close scrutiny concerns employees who request leave to care for a spouse, parent, or child with a serious health condition. Under the ADA, an employer is not required to allow a non-disabled employee to take leave to care for a spouse, child, or parent with a disability. Under the FMLA, however, an employer is required to provide up to 12 weeks of unpaid leave for such purposes. Therefore, the denial of a request to care for a spouse, parent, or child with a serious health condition would not violate the ADA but would violate the FMLA.

Medical Examinations, Medical Inquiries, and Certifications

The FMLA permits an employer to require its employees to provide written certification from a health care provider to verify the need for medical leave. The ADA, however, prohibits inquiring into whether an employee is an individual with a disability and into the nature and severity of the disability. Thus, if an employer requires an employee with a medical condition that is considered a disability under the ADA to provide a written certification from a health care provider to verify the need for medical leave, the employee may, in effect, be required to provide information concerning a disability or its nature and severity. While this requirement is permissible under the FMLA, it would violate the ADA. At the present time, there is no definitive answer to this problem. However, in an attempt to comply with the ADA and the FMLA, it is suggested that the following procedures be followed when verifying employee leave requirements under the FMLA:

- Written certifications allowed under the FMLA should be narrowly tailored to secure only information necessary to verify leave requests. For example, an employer may ask the physician to only verify that leave is necessary, and not to disclose the medical condition requiring the leave.
- If the employer requires a certification to support a request for medical leave, the information required should be job-related and consistent with business necessity (e.g., need for the leave). Seeking more information than necessary to verify the leave request may violate the ADA.
- Employers should not inquire into possible future effects of an employee's "serious health condition" during the certification process. For example, if a written certification verifies that an employee has cancer, the employer may not inquire into whether the employee's illness is terminal.
- Supervisors should be instructed not to discuss leave requests or medical conditions with employees. One person knowledgeable about leave policies and the ADA should be designated as the person responsible for processing leave requests.
- The employee's supervisor should only be told that the employee will be taking leave and will return at a specified date.

While these procedures might not prevent a violation of the ADA, they should be helpful in showing a good faith attempt to comply with the ADA in situations where an employee with a "disability" under the ADA is required to obtain a certification to verify a FMLA leave request.

Requiring medical certification from employees who request medical leave also raises some privacy issues under the ADA. Under the ADA, an employer must keep information concerning a disabled employee's medical condition confidential and such information must be maintained in separate medical files; access to these files must be restricted to those who have a need for that information. Thus, a written certification under the FMLA verifying an employee's leave request based on a condition that is a "disability" under the ADA must also be kept confidential to comply with the ADA.

Another concern arises when a female employee takes FMLA leave for the birth and care of a child. The Pregnancy Discrimination Act of 1978 requires an employer to hold open the job of an employee who is out on pregnancy-related leave on the same basis as the employer holds jobs open for employees on sick or disability leave for other reasons. This means that if an employer would not normally hold open a job for an employee on disability leave, it does not have to hold open a job for a woman on a pregnancy-related leave. Under the FMLA, however, a woman who is out on a pregnancy-related leave must be reinstated to her original position or to an equivalent position, regardless of the way the employer treats other "disabled" employees.

In complicated situations like those described above, it is suggested that you consult with your attorney prior to taking action.

Light Duty, FMLA, and Workers' Compensation

Another problem arises with employees who are on FMLA leave but are released by their health care provider to return to work in a light-duty position. Under the FMLA, an employee may not be forced to return to work in a "light-duty" position before his/her FMLA leave entitlement has expired, as forcing the employee to return to such a position would violate his/her right to be reinstated to the same or equivalent position.

This is true even if the employee is on FMLA leave as a result of a workers' compensation-covered injury. Under state workers' compensation statutes, an employee's workers' compensation benefits may be suspended if the employee refuses a light-duty assignment. Thus, if an employee is injured on the job and the injury also qualifies as a "serious health condition" under the FMLA, the employee will qualify for both workers' compensation benefits and FMLA leave. This would allow the employee to receive workers' compensation benefits and have his health insurance maintained under the FMLA. If the employee is offered a medically approved light-duty position, the employee may decline the position and exercise his/her FMLA rights and remain on FMLA leave. If the employee accepts the light-duty position in lieu of FMLA leave or returns to work before the 12 weeks are up, the employee still retains his/her right to reinstatement to his or her original or equivalent positions until 12 weeks have passed.

Another problem that arises with employees who are receiving workers' compensation benefits and are on FMLA leave concerns the substitution of paid leave for unpaid FMLA leave. In such cases, the employee must elect whether to receive paid leave or worker's compensation benefits. If the employee elects to receive worker's compensation benefits, the employer cannot require the employee to substitute paid vacation or other paid leave for unpaid FMLA leave. However, worker's compensation leave may be counted against the employee's 12-week FMLA leave entitlement provided the employer properly designates the leave as FMLA leave.

FMLA's Maintenance-of-Benefits Provision and COBRA Coverage

The FMLA requires employers to maintain health insurance or benefits while an employee is on FMLA leave. Many employers want to know how the maintenance-of-benefits provision of the FMLA affects COBRA coverage.

COBRA provides that a reduction in hours of employment that results in loss of insurance coverage is a qualifying event that entitles employees to elect up to 18 months of COBRA coverage. In most cases, a leave of absence is considered a reduction in hours and is a qualifying event under COBRA. The IRS, however, takes the position that leave under the FMLA is not a qualifying event as the employer is required to continue health insurance coverage during the period of an employee's FMLA leave. Thus, as FMLA leave is not a qualifying event, any insurance coverage provided during the FMLA leave is not COBRA coverage and cannot be credited toward the satisfaction of COBRA's requirements.

Under IRS guidelines, a "qualifying event" will occur for employees on FMLA leave if all three of the following conditions are met:

1. An employee, spouse, or dependent child is covered under the employer's group health on the day before the first day of the FMLA leave or becomes covered during the FMLA leave;
2. The employee does not return to employment with the employer at the end of the FMLA leave; and
3. The employee, spouse, or dependent child would, in the absence of COBRA coverage, lose coverage under the group plan before the end of their maximum coverage period under COBRA.

If the three above-described conditions are met and a qualifying event occurs, the date of the qualifying event for COBRA purposes is normally the last day of the FMLA leave, which is normally the last day of the 12-week FMLA leave period. However, the qualifying event could occur earlier if the employee informs the employer that he or she will not be returning to employment. The qualifying event could occur after the end of the FMLA leave period if the employer's insurance plan provides coverage beyond the FMLA leave period. For example, the insurance plan may provide coverage to the end of the month and the FMLA leave period ends prior to the end of the month. In such cases, the qualifying event occurs on the date insurance coverage is actually lost, which, in our example, is the end of the month.

ADA and Workers' Compensation

Employee injuries at work may raise ADA issues. Here are some guidelines for dealing with these issues:

- An employer may request a medical exam after an injury if the injury was job-related and consistent with business necessity. The employer may require a job-related medical exam, but not a full physical exam.
- An employer cannot refuse to allow a qualified person with a disability due to an injury at work to return to work because the worker is not fully recovered from the injury unless:

  - ▲ The employee cannot perform the essential functions of the job with or without an accommodation.
  - ▲ Returning to work would pose a significant risk of substantial harm that could not be reduced to an acceptable level with reasonable accommodation.

- Reasonable accommodation may include reassignment to a vacant position if qualified.
- Not every workers' compensation injury is an ADA disability, but you need to treat every worker's compensation injury as a potential ADA claim. Actively evaluate each injury to determine if the resulting impairment to the employee, if any, qualifies under the ADA definition of a disability protected under the law. If an injured worker requests to return to work, the employer should take that request seriously and reasonably accommodate any lingering impairments. Employer refusal to return an employee to work may result in that employee filing an ADA claim. After an employee returns to work, an employer may need to make further reasonable accommodations to help the employee perform his or her essential job functions.
- The employer is subject to further ADA claims if the employer passes over previously injured workers for promotion or lays them off because of injury.

■ No-fault absenteeism policies that fail to reasonably accommodate disabled workers violate the ADA. Employers are justified by business necessity to terminate an employee due to excessive absenteeism for time-critical functions or if the presence of the employee is critical to the job *and* no amount of reasonable accommodating will allow an employee to perform the essential job functions without undue hardship.

MAINTAINING A DRUG-FREE WORKPLACE

Introduction

You may be required to adopt a program designed to prevent unlawful drug usage among your employees. This program should include a policy against drug abuse in the workplace, a communication and education program about drug-abuse, and notification of drug abuse problems. However, it does not require drug testing.

As it is used in this section, the term "drugs" refers to controlled substances such as narcotics, cannabis, stimulants, depressants, and hallucinogens. Likewise, the abuse of lawfully obtained prescription drugs is also prohibited.

Even if you are not required by law to institute drug-free workplace practices, you may still wish to implement certain procedures for creating and maintaining a drug-free workplace. There are several important reasons for doing this:

■ Employees with drug-abuse problems are potential dangers to themselves, to co-workers, to customers or suppliers, and to your business. Providing a safe working environment is part of your general obligation as an employer.
■ Employees with drug-abuse problems tend to reduce your business' profitability through absences, lower productivity, and poor-quality work.

Legal or Regulatory Basis

The Drug Free Workplace Act of 1988.

Are You Covered?

This law applies to all federal government contractors with contracts of $25,000 or more, as well as to anyone who receives a federal grant.

Each state may have its own laws and regulations concerning drug abuse in the workplace. You should contact your state agency or department of labor, employment, or the like for information about any specific state requirements. Employees under treatment for drug abuse may be covered by federal, state, or local laws such as the ADA.

Action Steps

If you are covered by this law, you should take the following steps:

1. Clearly communicate to your employees that illegal drug abuse will not be tolerated. State and local laws may regulate this area. Publish the policy or a statement that notifies all your employees that the unlawful manufacture, distribution, dispensing, possession, and/or use of a controlled substance in the workplace is prohibited.

   The sample employee handbook in Section II includes a copy of a general drug-free workplace policy. Listed below are several ways to communicate this policy to employees:

   ■ Send every employee a copy of your drug-free workplace policy.
   ■ Post a copy of this policy on company bulletin boards.

■ Include this policy in your employee handbook.
■ Include in employee education or training programs (see item 3).

Be sure to apply your policy consistently.

2. Use drug testing from a certified lab to screen final candidates for job openings. Current illegal drug use is not a protected disability, which means you can test applicants at any time and reject applicants who test positive. Work with a reputable lab to set up and follow proper drug testing procedures. When you consider what a drug abuser may cost you, the investment in drug testing during candidate selection is one well taken.

3. Establish a drug-free workplace awareness program. This educational effort should cover the following items:

■ The dangers of drug abuse in the workplace.
■ A review of your company's policy on drug abuse (see the prior item).
■ Drug counseling or rehabilitation assistance programs.
■ How employees who violate this policy will be punished. You are required to impose a sanction on and/or require rehabilitation of any employee convicted by a court of law of such a crime.

You may be able to offer educational programs free or for a nominal charge through your medical insurer, local hospital or mental health agency, or private groups such as Alcoholics Anonymous.

4. All employees involved in work under a contract or grant must be given a copy of the policy with the proviso that they agree to the terms of the statement. They should also agree to tell you of any workplace-based, drug-related conviction within five days.

   Once you become aware of such a conviction, you must notify the contracting agency within 10 days.

Related Matters

The various laws covering disabilities may also apply here. What this means is that you should be careful before terminating someone you suspect or have found guilty of drug abuse.

Current illegal drug use is not protected under statute. Employees found illegally involved in drug abuse in the workplace can be fired.

However, an employee with a prior drug abuse problem or an employee currently suffering from alcohol-related problems would be considered disabled under the 1990 Americans With Disabilities Act. Under this act, employers of 15 or more employees are obligated to make an effort to reasonably accommodate the employee (e.g., by allowing time during the day to visit a clinic). Even so, you can still expect any employee to remain drug- or alcohol-free on the job and to meet the expectations of the job. Employees who are unfit for duty should be removed from the job.

For More Information

Contact the National Clearinghouse for Alcohol and Drug Information:

11426-28 Rockville Pike
Suite 200
Rockville, MD 20852

800-729-6686
website: www.health.org

Classifying Employees

JOB DESCRIPTIONS

Introduction

A job description is a summary of the important responsibilities and duties expected of a position in your organization. Although you are not legally required to have job descriptions, they can help you manage employees more efficiently and effectively.

How Job Descriptions Can Help You Be More Productive

There are several ways job descriptions can help you manage your human resources programs more effectively:

1. You can use job descriptions for completing wage and salary surveys so that you can price jobs competitively.
2. You can use job descriptions to inform applicants about the job and to guide the preparation of interview questions and procedures.
3. Job descriptions help communicate performance expectations and thus help you manage employee job performance.
4. Job descriptions focus evaluations of employee performance and are an important foundation for disciplinary action.

In addition, job descriptions that define the essential functions of the job give you an advantage when considering hiring applicants with disabilities.

Job Descriptions and the Americans With Disabilities Act (ADA)

The Americans With Disabilities Act of 1990 was passed to encourage and promote the hiring of qualified people with disabilities. A person has a disability if the physical or mental impairment limits major life activities (such as walking or seeing), the person has a record of such impairment (such as heart disease), or the person is regarded as having such an impairment (e.g., clinical depression).

In considering an applicant with a disability for a job, you should determine whether the applicant can perform the "essential functions" of the job. This consideration may include determining if the person can perform those functions with "reasonable accommodation."

As an employer, you can protect yourself against claims of discrimination if you have written job descriptions that list the essential functions of a position. Such a listing is a natural ingredient of job descriptions.

Are You Covered by the ADA?

Employers who have 15 or more full-time employees are covered by the ADA.

What Should a Job Description Cover?

Typically, a job description will include the following descriptive information about a job:

- Identifiers: job title, department, reporting relationship, exempt status.
- A summary of the job's purpose
- A listing of job responsibilities, with the essential functions of the position noted.
- A statement of any supervisory responsibilities.
- Required minimum qualifications (knowledge, skills, experience).
- Any specific or unusual physical demands or working conditions.
- Any hazardous working conditions.

A sample job description form appears on page 131 of Section III.

How Is a Job Description Created?

Typically, when you write a job description for the first time, you must get information about the position from the supervisor and/or employees in the job. Other people (customers, suppliers, coworkers) might also be consulted.

The following guidelines can help you complete the key sections of a job description.

Identifiers

- Use a title that identifies the job's real purpose. Do not inflate a title; avoid qualifiers like Assistant 11 or Junior Bookkeeper.
- Identify whether the position is exempt or non-exempt. (See "Exempt or Non-Exempt Employee" later in this section.)
- Identify the supervisor of this position.

Summary

- Write one to three sentences that describe the position's primary purpose. Why does the job exist? What is it primarily designed to accomplish?

Job Responsibilities

- List the primary duties performed in the position. Identify these duties as the position's essential functions.
- Be sure to list important tasks that may be performed only occasionally or irregularly.
- Begin duty statements with action words (calculates, measures, etc.).
- If you are covered by the ADA, be sure to provide a complete listing of all essential job functions to document what the job entails.

Supervisory Responsibilities

- Indicate how many people this position supervises (if any), and what those positions are. Because of the potential liabilities in the areas of overtime and sexual harassment, a position should only be given a supervisory designation after careful thought.

Qualifications

- State the minimum and actual qualifications a person must have to perform the job. The qualifications must be job-related.

Unusual Demands or Conditions

- Identify any distinct lifting, force, movement, or endurance capabilities required of the position.
- Note any unusual environmental or working conditions, such as excessive heat, hazardous situations, hazardous chemicals, or the potential for irregular working hours.

Disclaimers

At the end of each job description, be sure to include appropriate disclaimers, such as:

- A brief at-will employment statement that the job description is not an express or implied contract.
- Management reserves the right to add, delete, or modify the job description.
- Other duties may be assigned by the supervisor and/or his or her designee.

Resources

Use the sample job description, physical and mental job demand assessment, and working conditions and environment assessment forms in Section III on pages 131, 133, and 134.

You don't have to write new job descriptions, at least from scratch. *Job Descriptions for the Home Building Industry, Third Edition,* contains a variety of tools to help you spell out your employee expectations right up front. It contains 40 sample job descriptions, plus information on job analysis, interviewing, and performance evaluation. The book includes a CD, too, so you can easily adapt forms and job descriptions for your company. Call 800-223-2665 or go to www.builderbooks.com to order it online.

INDEPENDENT CONTRACTOR OR EMPLOYEE?

Introduction

A person can work for you as either an employee or an independent contractor. Depending upon that person's working status, you have different obligations to the worker and to governmental taxing authorities. Furthermore, an employee enjoys certain rights in the relationship that are not available to independent contractors.

If you are using someone as an independent contractor, you should be sure that person meets the criteria of an independent contractor. Just because you call the individual an independent contractor does not make the person an independent contractor under the law. You may be held liable for back employment taxes and penalties by federal regulatory agencies if you inaccurately classify an employee as an independent contractor.

Background

As an employer, you are required to collect various payroll taxes from employees when you pay them. You then pay these taxes to the various taxing authorities throughout the year. You don't collect income taxes from independent contractors.

At the federal level, the IRS has become increasingly alarmed by the faulty use of independent contractor status as a way to avoid collecting and paying taxes. The IRS is clamping down on employers who abuse the independent contractor classification.

What This Means for Your Business

The burden of proof rests on you to demonstrate that the person is in fact serving in an independent contractor's capacity rather than that of an employee. Otherwise, it is in your best interest to make sure you treat that person as an employee.

What Is an Employee?

In general, you have an employee working for you if you:

- Set the hours of work.
- Determine how the tasks will be done.
- Pay an hourly wage.
- Supply materials and tools.
- Provide space, communication, and/or support services.
- Critique the actual work done.

General Guidelines

If you use independent contractors, you should make sure those people are in fact working in an independent contractor relationship. It is prudent to keep a written record of the reasons of this determination.

In general, an independent contractor is someone who:

- Retains control over the nature of the work done, and exercises independent initiative and judgment, including when and how the job is completed.
- Works on a temporary basis or per-job basis.
- Owns and supplies his or her own tools and equipment.
- Operates on a profit and loss basis (not paid for time solely).
- Maintains his or her own insurance for liability as well as workers' compensation insurance.

IRS Guidelines

The IRS uses the following list of 20 questions to determine if a position is that of an independent contractor or an employee. Answering "yes" to many of these questions indicates the position is being performed by an employee.

1. Is the worker required to comply with your instructions about when, where, and how the work is to be done?
2. Is the worker trained to do the work in a particular way?
3. Does the worker do work that is provided as part of your business' regular operations?
4. Do you expect the work to be done by the individual personally?
5. Do you pay others to help the worker perform the services under contract?
6. Is the working relationship a continuing one?
7. Do you establish the hours of work?
8. Is the person required to work full-time for your business?
9. Does the person only work at your business location(s)?
10. Do you direct the work being done?

11. Is the person required to give you regular reports on job performance?
12. Do you pay the person a regular wage or salary?
13. Do you reimburse traveling or other business expenses?
14. Do you provide the tools and materials?
15. Does the person not have a significant investment in the facilities used to do the work?
16. Does the person not realize a profit or loss?
17. Does the worker work exclusively for you?
18. Does the worker not make his or her services available to other businesses while still working for you?
19. Can this person be dismissed for reasons other than non-performance of contract duties?
20. Can the worker quit at any time without incurring a liability for failure to complete the job?

Note: *Consider* all *of the above guidelines and criteria when you are trying to determine whether a worker is an employee or an independent contractor. Don't use just one or two factors to make your decision.*

Obtaining Assistance

If you are unsure about whether a position is that of an employee or an independent contractor, the IRS will advise you on the matter. Simply complete the SS-8 Determination of Worker Status form available in English and Spanish on pages 135 and 140 of Section III and submit it to the IRS, which will make the determination and notify you of its decision. To download a copy of the forms online, go to http://www.irs.ustreas.gov/formspubs/index.html.

Other Things to Do

■ If you use a worker as an independent contractor, it is prudent to have a signed contractual understanding to that effect. Often, a contractor will provide a written agreement on the terms and conditions of the work to be done. For sample contract language related to independent contractors or trade contractors, see *Contracts and Liability for Builders and Remodelers, 5th Edition,* available from BuilderBooks. Go to www.builderbooks.com to order it online.

■ Make sure managers know not to exert undue control or direction over contractors' performance on the job. Managers should not issue disciplinary action to contractors.

■ If you're seeking an independent contractor, make sure you mention this in "help wanted" ads and other communications related to the position.

■ If you must change a worker's status from independent contractor to employee, try to do so on January 1. That way you can issue 1099 forms for the prior year and W-2 forms for the new year.

■ You are not responsible for collecting income or social security taxes on amounts you pay to independent contractors. However, if you pay an independent contractor $600 or more in a year, you must issue and file a form 1099-MISC reporting the amount paid.

■ Ensure that independent contractors carry their own workers' compensation insurance and provide you with an insurance certificate.

Penalties for Misclassifying

If you are found to have unreasonably misclassified an employee as an independent contractor, you are liable for paying back taxes due and can be held personally liable for fines and penalties.

For More Information

The IRS supplies a variety of brochures at no cost. A good general brochure to consult at the beginning is "Publication 15, Circular E, Employer's Tax Guide." Go to www.irs.ustreas.gov/formspubs/index.html to download copies of forms and publications online.

EXEMPT OR NON-EXEMPT EMPLOYEE?

Introduction

The Fair Labor Standards Act (FLSA) defines the basic rules for employee compensation. In particular, the FLSA sets requirements for minimum wage and overtime pay. These rules apply to all employees unless the employee works in a position that is "exempt" from these requirements. Employees who are covered by the FLSA and who must be paid a minimum wage and overtime are "non-exempt" employees. Employees in these positions are also often called "hourly" employees.

Note: *If you do not pay a non-exempt employee correctly, you are liable for back pay and possible penalties. Therefore, you should be sure that you accurately classify an employee as "exempt."*

Are You Covered?

The FLSA covers almost all employers. All employers of certain enterprises having workers engaged in interstate commerce; producing goods for interstate commerce; or handling, selling, or otherwise working on goods or materials that have been moved in or produced for such commerce by any person are covered by the FLSA. The term "enterprise" includes, among many other things, businesses engaged in construction or reconstruction and businesses that have an annual gross volume of sales of $500,000 or more.

Basic Rules of the FLSA

In general, under the FLSA, covered employers are required to:

- Establish a work week, beginning on the day and time of your choice, which then covers the remaining seven consecutive days.
- Pay their employees a minimum wage ($5.15 as of date of publication). Many states also have minimum wage laws. Where an employee is subject to both the state and federal minimum wage laws, the employee is entitled to the higher of the two minimum wages.
- Pay an overtime rate equal to at least 1.5 times the usual rate of pay to employees who work more than 40 hours during the work week.
- Limit how much and what kind of work minors may perform.
- Keep records of wages, hours, and related items; the form can be the employer's choice.

What Is an Exempt Employee?

Employees in certain positions may be exempted from the minimum wage and overtime requirements of the FLSA. These positions are called "exempt" positions. In general, there are four kinds of exempt positions:

1. Executive
2. Administrative
3. Professional
4. Outside sales

Tests for Exempt Positions

There are very specific rules that determine whether a position is exempt or not. Given the penalties that can be imposed on you if you do not pay a person properly, you should make sure that a position in fact qualifies as exempt.

You'll find a form containing tests for exemption under the FLSA on page 145 of Section III.

An outside sales position is exempt if the employee spends at least 80 percent of his or her time each week selling or obtaining orders for goods or services and this work is usually done away from the employer's place of business. There is no salary requirement.

Action Steps

If you are covered by the FLSA, you should be sure to do the following:

- Begin by assuming that the employees working for you are non-exempt and making sure they are paid overtime for hours worked in excess of 40 hours during the work week.
- If you believe that a position is "exempt," fill out a copy of the test form for exemption under the FLSA on page 145 of Section III. Be objective and impartial, and remember you are rating the duties and responsibilities of the position—not the person in it.
- If you determine the position is exempt, notify the employee and explain the decision. In particular, the individual should be told that the position does not earn overtime payment. Also, keep a copy of the test for future reference and justification.

Compensation and Benefits

COMPENSATION PROGRAM AND ADMINISTRATION

Introduction

Your compensation program is a major concern for both you and your employees. For you, it is a major expense that involves lots of planning and preparation. Your employees' livelihood depends on receiving their salaries or wages correctly and on time. Errors in paying your employees correctly may erode or destroy employee goodwill and/or invite regulatory or legal action.

Legal or Regulatory Considerations

Here are some basic rules for compensating employees and the laws or regulations that support them:

Issue	Rule	Law or Regulation
How often to pay employees	The Fair Labor Standards Act requires that wages for non-exempt employees be calculated on a weekly basis. However, state laws regulate the frequency with which non-exempt employees are to be paid; this usually is not more often than bimonthly or every two weeks. Exempt employees can be paid on a monthly basis.	Fair Labor Standards Act State laws
Compensating men and women	You cannot discriminate by compensating a woman less than a man if the woman is doing a job that is substantially the same as a job done by a man. Jobs are substantially the same if they require the same skill, effort, and responsibility. Some differences in pay rates are permissible if they are due to performance factors such as production bonuses or merit increases.	Equal Pay Act

Issue	Rule	Law or Regulation
How much to pay employees: base compensation	In general, employees should be paid at least the minimum wage of $5.15 per hour (as of 2003). Exceptions: ■ Exempt employees who receive a "training wage." ■ If you are working on a federal or federally financed construction project worth $10,000 or more, mechanics and laborers must be paid at least the minimum wage rate set by the Secretary of Labor for particular industries.	Fair Labor Standards Act The Walsh-Healey Act
How much to pay employees: overtime	In general, non-exempt employees who work more than 40 hours during the work week should be paid at 1.5 times their hourly rate for all hours over 40. Exempt employees do not have to be paid overtime.	Fair Labor Standards Act
"Comp" time as a substitute for overtime	You cannot substitute "comp" time off for overtime wages due. That is, if a non-exempt employee works more than 40 hours during the work week, that employee must be paid the overtime rate for the hours worked over 40. A non-exempt employee may not make an agreement to substitute compensatory time off for overtime hours worked.	Fair Labor Standards Act
Paying for travel to and from work	In general, an employer does not have to pay for an employee's ordinary time spent traveling to and from work. However, unusual or special assignments that require travel time beyond the ordinary (e.g., to attend a meeting in another town) may need to be compensated.	Portal to Portal Act
Paying for preparation or cleanup	You may need to compensate an employee for time spent in preparation or clean up under the following conditions: ■ Such time is customary for the job. ■ The time is significant, not incidental (15 minutes or more). ■ It must occur immediately prior to or after the primary work done; that is, the activities are indispensable to task performance. ■ Such time is requested by the employer.	Fair Labor Standards Act

(Continued)

Issue	Rule	Law or Regulation
Paying for breaks	Rest periods of short duration are common and must be compensated. Rest periods when the employee is removed from the job for 20 minutes or longer need not be compensated.	Various court judgments
	Meal breaks are not considered compensable time and do not need to be paid for under these conditions:	
	■ The employee is completely removed from job duty. ■ The purpose of removal from duty is eating a meal. ■ The break is at least 30 minutes.	
Paying for leave	In general, an employer is not required to pay employees for any time off granted to them. This includes time off for vacation, sick leave, holidays, or personal reasons.	No statute
	However, if you have a policy, it should be applied consistently among all employees. Develop a leave policy you can live with and can apply consistently.	
Docking pay	Docking the pay of an exempt employee for time taken off during the week (for any reason, including medical appointments, personal days, etc.) will lead to a loss of that position's exempt status and you will be liable for overtime payments. In general, the best policy is not to reduce an exempt employee's salary for any absence from the job for less than a full workday.	Recent court decisions
	A staff member may have his or her salary reduced for unpaid FMLA-qualified leave without causing the employee to lose FMLA exemption status, including deductions for intermittent and reduced leave schedules. However, deductions taken for unpaid leave before the employee is eligible for FMLA or for a reason not qualifying under FMLA affects employee exemption eligibility.	

Issue	Rule	Law or Regulation
Wage garnishments	There are limits on how much can be deducted from an employee's paycheck because of a judgment against the employee. Be sure to obtain a court order before processing the deduction. Consult your state regulations. Some states do not allow any adverse actions to be taken against employees who have one-wage garnishments.	Wage Garnishment Act
Taxes on bonuses and awards	Cash bonuses and awards must have income tax withheld. This is most easily done by combining the bonus or award with regular pay and computing the amount to withhold on that total.	Internal Revenue Code
Employment tax payments	You are required to collect income and social security taxes and then forward those amounts to the IRS and any applicable state authorities on a timely basis. Consult the IRS and/or your state authority for specific timing and procedures.	Internal Revenue Code

The laws of your state may also affect how you pay employees. Again, consult your appropriate state authority for more information and details.

Wage and Salary Administration

It's best to establish and use a wage and salary administration program that allows you to produce optimum results in the most cost-effective manner.

Keep in mind that the cheapest wage administration plan may not be the best one for your business. Some slight increases in compensation levels or in how wages are earned may yield much better returns to your business by increasing employee satisfaction levels.

Action Steps

1. Provide competitive compensation. A competitive compensation package pays close to what the market pays people for doing a given job with the employee's qualifications and skills.

   If you don't provide a competitive wage, you expose your company to various risks. First, by paying under the market, you increase turnover. This boosts your business costs from lost and poor productivity as well as paying for repeat employee searches. Second, by overpaying employees, you reduce profitability.

   One way to make sure you provide a competitive compensation program is to compare your employees' salaries and wages to data from salary survey reviews. Salary surveys are systematic efforts to determine what different jobs are paid and receive in benefits in a labor market. Here are several sources for salary survey information:

   - The local chapter of the Society for Human Resource Management (SHRM) may sponsor and/or conduct periodic salary surveys of jobs in your area. The national headquarters can give you a local contact. The SHRM phone number is 800-283-SHRM. Visit the organization's website at www.shrm.org.

- Trade associations, chambers of commerce, or other groups may have competitive compensation information. Trade publications such as *BUILDER* magazine and *Professional Builder* magazine conduct their own surveys and publish the results. In addition, the Bureau of Labor Statistics compiles and publishes general wage rate information for different occupations in different localities. Contact the Bureau of Labor Statistics for details at 202-691-5200. Visit its website at www.bls.gov.
- Check local help-wanted ads to see what wages other employers in your area offer.
- Conduct your own salary survey. Here are some tips:

  ▲ First, identify whom you want to participate in the survey. The obvious candidates are owners of businesses that compete for the same kinds of employees that you do. While this would certainly include other builders, you may also want to include building supply vendors, trade contractors, etc.
  ▲ Offer to share the results. You may be able to encourage more participation by sending a summary of the results to those who provide you with information.
  ▲ Create a simple survey form. The form should include an introductory statement explaining who you are, what you are asking for, and why you are seeking the information, along with a short, easy-to-fill-in survey that asks for the following information:

    ● Title of each position (or a certain job, if you only want to know how much other employers pay their supervisors, for example)
    ● Brief description of key duties
    ● Pay range for the position
    ● Wage currently paid to employee (hourly rate or annual salary)
    ● Incremental wage increases/date(s) of such increases
    ● Employees' credentials
    ● Benefits offered

  ▲ Compile the results. Instead of calculating average wage rates, many builders prefer to use median rates. Simply rank order all the results from lowest to highest, and then use the middle number. This approach allows you to avoid the biasing effects of one very high or low wage rate.

2. Implement incentive compensation. An incentive compensation plan allows you to pay employees based on what they do and how well they do it. Keep in mind that traditional hourly compensation simply pays people for the time they are on the job. There are several kinds of incentive compensation plans that might make sense for your business:

- A profit-sharing plan. Set an annual profit goal for your business. If profits for the year ended up 10 percent higher than planned, for example, then you'd share the profits with all your employees. You might pay a flat bonus amount, such as $250 per employee, or a variable amount. The variable amount may be computed in the following way:

  ▲ For this example, assume that the planned profit for the year was $100,000, and you ended up with a profit of $110,000, or a $10,000 (or 10 percent) gain over plan.
  ▲ Plan to set aside some portion of that additional profit. Certain plans recommend a 50 percent sharing of profit. Here, that would amount to $5,000.
  ▲ Divide the portion of the profits to be shared ($5,000) among your employees. Assume you have three full-time employees who together earn compensation of $85,000 (they each earn

$35,000, $30,000, and $20,000, respectively). Proportionately, the employees earned 41 percent (35/85), 35 percent (30/85), and 24 percent (20/85) of the total compensation amount. You could pay each employee a bonus in proportion to their compensation. Thus, the first employee would receive a bonus of $2,050 (5,000 × .41), and so on.

- A merit performance plan. Under this plan, employees receive bonuses based on how well they performed their jobs over the past year. Employees who performed very well might receive 6 percent bonuses, while employees who performed adequately might receive 3 percent bonuses.

   A merit performance requires an objective performance evaluation procedure. (See the information on "Performance Appraisal Systems" later in this section.)

3. Monitor employee opinions of your compensation system. You want employees to believe that you are paying them fairly. Morale and/or discrimination problems can crop up if employees believe you are acting in an arbitrary or illegal fashion. For example, you should avoid overcompensating a "complaining employee," which could be viewed as inconsistent compensation. Here are several steps you can take to increase employee perceptions of fair treatment:

- Implement compensation consistently and fairly among all of your employees.
- Let employees know you conduct or participate in salary surveys. If possible, let them see the results.
- Indicate your willingness to discuss and explain compensation plans with employees if they have any questions.
- Create and publish clear, specific rules for how employees' compensation will be changed. If you use an incentive plan, be sure that you lay out the rules in advance so that employees know what they are. Be sure you stick to the rules and apply them unilaterally
- Conduct an employee survey to find out what your employees want in a compensation plan. Employees may want non-monetary benefits like child-care services, flex-time working schedules, etc.

For More Information

Consult the following sources if you have questions:

For minimum wage and overtime:
U.S. Department of Labor
Wage and Hour Division
www.dol.gov/esa/whd/

For pay periods:
Your State Department for Employment

For employment taxes:
The Internal Revenue Service
www.irs.gov

Your State Revenue Department

PAYROLL

Introduction

You should establish the following routine administrative procedures for your payroll system:

■ **Withholding allowances**—New employees should complete a W-4 form authorizing the number of deductions. This should be done on the employee's first day on the job. The deductions are in effect until the employee files a new form, which should be done whenever the employee's status changes.

■ **Calculating wage rates**—You may pay employees on any of the following schedules: hourly, fixed weekly, fixed wages for some other period, commissions, day rates, or some combination. Make sure the rate you pay equals or exceeds the minimum wage; if you do not pay an hourly rate, convert your pay into an hourly rate to ensure that you meet or exceed federal and state minimum wage rates.

Calculate the hourly rate by totaling the employee's work week compensation, including salary, commissions, bonuses, incentive payments, shift differentials, and "dirty work" premiums. Then divide the total by the number of hours worked during the work week.

■ **Making deductions**—You are allowed to make certain deductions from payroll for taxes, meals or facilities provided, uniforms, voluntary wage assignments for loans or advances, garnishments, and so on. If there is an indebtedness created with the employee, you should use a voluntary form that indicates the debt created and authorizes payroll deductions. Employees must authorize payroll deductions other than federal and state employment taxes and garnishments. See the sample payroll deduction authorization form on page 146 of Section III.

■ **Calculating overtime pay**—You only pay overtime for time actually worked. Paid time for sick, holiday, vacation, or other leave is not included in the hours worked per week for overtime calculations. For example, if an employee worked 35 hours in four days during the week and also had a paid holiday of eight hours, the employee is not paid overtime for three hours.

The overtime rate should be at least one-and-one-half times the employee's regular hourly wage. However, that regular hourly wage is the actual compensation typically received, not simply any fixed routine wage. For example, if an employee is paid a fixed wage of $6.50 per hour but also receives a productivity bonus equivalent to $1.00 per hour, the overtime rate of pay is 1.5 times the actual $7.50 per hour rate ($6.50 + $1.00). In this example, the overtime rate would be $11.25 per hour ($7.50 × 1.5), not $9.75 per hour ($6.5 × 1.5).

The kinds of supplemental and variable pay that must be included in calculating overtime pay rates are:

▲ Piece work
▲ Cost-of-living bonus
▲ Shift premiums
▲ Production bonus
▲ Sales commissions
▲ Merit increases
▲ Productivity bonuses

To figure the actual rate of pay, simply add in the amount earned through the supplemental or variable pay plan to the existing base pay rate. Divide that amount by the hours worked during the week to produce the adjusted rate. Many payroll processing systems automatically calculate the actual rate of overtime pay.

You do not have to include lump-sum awards of a discretionary nature based on performance improvement suggestions, paid leave from work, expense reimbursements, supper money, or employee referral bonuses in the calculation of the proper overtime rate.

■ **Maintaining payroll records**—You are required to keep records of wages, hours, and related payroll matters. No specific format is required, but the records you maintain should include the following information:

▲ Personal information about the employee (name, address, age, social security number)
▲ Total hours worked each day and for the work week
▲ Total daily or weekly straight-time earnings
▲ Regular hourly pay rate (the straight-time rate of pay)
▲ Total overtime pay for the work week
▲ Deductions from or additions to wages
▲ Total wages paid each pay period
▲ Date of payment and pay period covered

■ **Providing W-2 forms**—You must provide copies of form W-2 to each employee from whom income, social security, or Medicare taxes have been withheld during the year. It must be provided to the employee no later than January 31 of the following year.
■ **Keeping tax records**—You should keep the following tax records for a minimum of four years after the due date. These records should contain your employer identification number, copies of the returns, and dates and amounts of deposits.

▲ For income tax withholding:

● Each employee's name, address, and social security number.
● The total amount and date of each wage payment and the pay period covered.
● The amount subject to withholding for each wage payment.
● The amount of tax withheld.
● W4 forms for each employee.
● Records of any voluntary, additional tax withholdings.

▲ For social security and Medicare taxes:

● The amount of each wage payment subject to social security tax and/or Medicare tax.
● The amount of social security and/or Medicare taxes withheld.

▲ For federal unemployment taxes:

● The total amount paid to each employee during the calendar year.
● The amount you paid into the state unemployment fund.

WORKERS' COMPENSATION

Introduction

Virtually all employers are required to provide a form of workers' compensation coverage to their employees. This benefit provides protection to workers who are injured or suffer an occupational disease arising out of the course of employment. This benefit is primarily regulated by state laws, so consult your state agency for specific details.

Insuring the Benefit

There are three main options for insuring workers' compensation:

- Through an authorized insurance carrier.
- Through a state-based insurance fund.
- Through self-insurance (if one meets financial qualifications and is approved by the state agency).

Action Steps

1. If your state requires workers' compensation, make sure your business is properly participating. Consult with your state agency for details.
2. Manage the workers' compensation program to minimize costs and encourage employees to return to work as quickly as possible. Options for improving the management of your workers' compensation program include:

   - Reporting claims to your insurer promptly; if you suspect an employee is abusing this benefit, report it to the insurer as soon as possible.
   - Getting the employee back to productive work as quickly as possible with some kind of light-duty assignment.
   - Adopting a good safety management program to reduce risks and hazards.
   - Keeping morale high, which encourages safety and a motivation to return to work quickly.
   - Training employees in proper safety and health procedures.
   - Providing appropriate protection and injury-prevention devices.

For More Information

Consult *Reducing Your Workers' Comp Costs: Strategies for Builders and Contractors,* which is available from BuilderBooks. Call 800-223-2665 or go to www.builderbooks.com to order it online.

UNEMPLOYMENT INSURANCE

Introduction

Unemployment insurance was instituted to provide some protection to employees who lose their job through no fault of their own. Employers (and in limited cases, employees) fund the insurance pool through unemployment tax payments.

Are You Covered?

Virtually all employers are covered by the Federal Unemployment Tax Act. You are subject to unemployment tax on wages paid to any employee if wages paid to employees totaled more than $1,500 during any calendar quarter of the year.

How It Works

Unemployment insurance is administered by the federal and state government. The current basic rate for Federal Unemployment Tax (commonly known as FUI or FUTA) is 0.8 percent of the first $7,000 of compensation paid to each employee. This rate may be adjusted upward if you work in more than one state or have not made timely payments of your state unemployment tax.

In addition, unemployment tax is also paid to the state in which your employees are working. The state rates and maximum compensation are established by the State Unemployment Agency and are usually based on your unemployment claim experience.

An ex-employee files for benefits with a state agency. If he or she qualifies, the ex-employee will receive a monthly payment for a certain period of time or until a job is found.

Consult your accountant for more information on rates in your area.

Action Steps

1. Make your tax payment on a timely basis. Otherwise, you may lose some of your state offsetting tax credits.
2. Be careful about laying people off too quickly, especially if there is just a temporary lull or downturn. Consider using paid or unpaid vacation time for staffing buffers. Also consider helping ex-employees find other jobs easily.
3. Contest the claims of any ex-employee whom you believe should not be given the benefit. In general, employees who voluntarily quit or are released for misconduct are ineligible for unemployment benefits (or, if deemed eligible, will have their benefits reduced by a penalty amount). Keep specific records that explain why each employee left employment and documentation regarding any performance problems or warnings. For misconduct firings, be sure you have documents of the performance basis for the termination.
4. Make sure statements made to state agencies are the same as the reasons told to the employee who was terminated or told to other governmental agencies or future prospective employers (if you release such information).
5. Examine all notices and forms to make sure they are correct before approving payments. Employees who are receiving severance or other kinds of payroll continuations may not be eligible to receive unemployment benefits until the severance period and/or payroll continuation ends. Watch for dual payments and report them to the unemployment insurance agency in your state.

For More Information

Contact your State Department of Unemployment Insurance.

Benefits Administration

EMPLOYEE BENEFITS MANAGEMENT

Introduction

An employer is required to provide two kinds of benefits:

- Workers' compensation
- Unemployment insurance

Beyond that, the employer can decide what kinds of benefits—if any—he or she wants to provide. In many markets, employers need to offer more benefits to attract the caliber of employees necessary to do the required work. The need to increase the level of benefits is a function of the labor market, not of government regulatory requirements. An effective benefits administration program involves understanding what benefits are typically provided in your labor market so you can offer a competitive benefits package.

Employee benefits can represent a significant portion of your total compensation and employee overhead expense. Yet many employees don't know about—much less appreciate—the benefits their employers offer. Even though you are not legally required to provide much in the way of benefits, you may be forced to do so by your local labor market to attract and retain quality employees. A well-designed benefits program can help you develop a strong and consistent labor force.

Classification of Employee Benefits

Employee benefits are traditionally grouped into the following categories:

- Pay for time not worked (vacation, sick, or other leave)
- Employee income and welfare protection plans

  - ▲ Legally required (unemployment/worker's compensation)
  - ▲ Voluntary (medical, disability, life insurance, retirement)

- Perquisites and services

Legal or Regulatory Considerations

In general, you are not legally required to provide much in the way of benefits. However, if you do provide benefits voluntarily, certain laws and regulations govern the way you provide them.

The main legal area of consideration is the Employee Retirement Income Security Act (ERISA). ERISA aims to ensure that employees have adequate information about the benefits they are provided and that benefits are provided fairly and equitably. ERISA is discussed further in "Employee Welfare Plans" in this section.

Benefits are generally not taxable when they are provided to employees. In some cases, benefits may create tax advantages for the employee and the employer.

Given the complexity of this field, you should consult with your accountant, banker, insurance representative, and/or financial or management consultant specializing in benefits before changing or revising your benefits program. This will help you better understand the options available to you and any potential costs or tax consequences.

Action Steps

1. Provide competitive benefits. As noted in the preceding section on wage and salary administration, one of the main ways to provide competitive benefits is to know what your labor market competitors provide. You can find competitive information about benefits offered in various ways as previously described, including your own survey.
2. Use your employees' input when you evaluate and/or plan for benefits. Make sure you ask your employees what they want in a benefits package. If you are thinking about or planning for benefits, knowing what employees value will help you make good decisions.
3. Build employee goodwill. Provide written summaries that outline the benefits; many of these summaries are required by the law. Obtain information from your insurer about the programs offered and the summaries required under ERISA.
4. Provide comparative information. If you learn what the market is offering in benefits, you may wish to tell employees what the typical benefits package is. This comparison is especially helpful if your benefits package is as good as—or better than—the market.
5. Control costs. One of the most costly benefits is medical insurance and related sick leave and/or workers' compensation. You should try to find ways to help control expenses in these areas. Here are some methods to help limit and control expenses in this area:

   ■ If you adopt a paid sick leave program, set it up so that it does not encourage abuse. For example:

      ▲ Allow unused sick leave to accumulate up to a maximum amount (say, 180 days or six months), at which time a long-term disability program may kick in. Tell the employees that each sick day is paid at 100 percent of regular earnings and that the more days accumulated, the greater their protection against loss of income from illness or injury.
      ▲ Create quarterly lotteries so that everyone with perfect attendance during the quarter can enter a drawing for some award of value (cash, day off, tickets to an event, etc.).
      ▲ Grant some additional amount of paid leave for off-season periods to employees who did not use sick leave during the year. Thus, you could give three days of additional paid leave during the winter for perfect attendance during the summer. This is a strong incentive for employees not to take sick leave when it is not needed.
      ▲ If you pay for accumulated, unused sick leave at termination, you can use a buy-back program at the end of each year or at termination. For example, you could pay 25 cents on each dollar of accrued sick leave. Say an employee has 10 days of unused sick leave. At an hourly rate of $8 per hour, that amounts to a total dollar value of $640.00 (10 days × 8 hours/day × $8 per hour). Twenty-five percent of that total would be a bonus of $160.

- Institute cost saving measures for your medical insurance. Work with your insurer to look at whether the items covered are necessary and/or properly structured. This may include avoiding questionable benefits (such as unrestricted psychiatric care), using outpatient services whenever possible, requiring second reviews and opinions on major operations, and requiring employees to make co-payments for medical and dental treatments. In addition, consider:

  ▲ Selecting lower cost health care providers, such as HMOs
  ▲ Encouraging employee wellness through such activities as stop-smoking programs, seminars on proper diet and nutrition or exercise, subsidized gym memberships, and free health screenings.

PAY FOR TIME NOT WORKED

Introduction

In general, you are not required by law to provide any pay for time not worked. The major exception is paying for short breaks of 20 minutes or less.

Paid leave benefits can be an important component of an employee benefits program. Even though you may not be legally required to provide such benefits, you may still want to provide various forms of paid leave as a way to attract, motivate, and retain a skilled workforce.

Vacation Leave

Employers are *not* required to provide a vacation benefit. There is no federal law that governs vacation leave, but there are a variety of state laws that may apply.

Employers sometimes wonder whether or not they have to pay for unused vacation as compensation when an employee terminates his or her employment. In general, if you allow employees to accrue or accumulate vacation time, they should be paid for any unused hours on termination. Consult your state employment agency for more details.

Holiday or Weekend Pay

Employers are *not* required to provide time off for holidays or on weekends, nor are they required to pay a premium rate to employees working on those days. However, you should try to make a reasonable accommodation for an employee's religious beliefs.

Some states do regulate how many days during the week an employee may work, and this can affect your weekend scheduling. For example, employees may be required to have two consecutive days of rest during a work week.

Sick Leave

Employers are *not* required to compensate employees for time missed from work because of illness. Neither are employers required to pay employees for accrued but unused sick leave on the employee's termination. However, if the employer has the policy or practice of doing so, then all employees should be paid for accrued but unused sick leave.

Personal Leave

Employers are *not* required to provide time off for personal reasons, nor are they required to pay their employees for any granted time off.

Leave for Jury Duty

Under the 1978 Jury System Improvement Act, you cannot discharge or otherwise discriminate against an employee who performs jury duty in a federal court. State laws may also govern this matter, also, so be sure to consult with the proper state authority for details.

Federal law does not require you to pay an employee for time spent on jury duty. State laws may differ on this matter, though.

As an employer, you may request that an employee be excused from jury duty if the employee's absence at that time would seriously interrupt your company's operations.

Voting Leave

Employers are not required by federal law to grant time off to vote. Many states, however, have laws governing this matter and require that employers provide time off for the opportunity to vote in state elections. Consult your state authority for details.

Sample Leave Policies

Consult the "Leaves and Absences" section of the sample employee handbook in Section II for policies covering various types of leave.

EMPLOYEE WELFARE PLANS

Introduction

As previously noted, with few exceptions, you are not required to provide any benefits to your employees. However, from a competitive position, you may need to provide a benefits package that attracts and retains a quality workforce.

Types of Employee Welfare Benefits

Employers typically offer some form of employee benefits in the following areas:

- Health or medical insurance
- Group life insurance
- Long-term disability insurance
- Retirement or pension plans

Some recent innovations to benefits plans include various forms of child- or dependent-care plans (sometimes offered on a tax-advantaged basis), prepaid legal assistance, and employee wellness programs.

Insurance or Trust Forms of the Plans

Typically, the employer purchases these benefits in the form of an insurance policy or a trust account. Self-insurance is an option, but should be undertaken only after careful thought and preparation.

It is essential to work with a qualified expert while developing an employee welfare plan. It may cost you more to do so, but it's worth the effort and expense to avoid problems and develop a plan that works best for you.

Legal and Other Requirements

Although you are not required to offer employee welfare plans, if you do so, you are likely to be covered by various laws and regulations regarding the way you provide the plans. Furthermore, in some cases (such as Simplified Employee Pension Plans), the insurance company or financial service firm from which you obtain the plan may have its own rules about the plan.

General Legal Rules

In general, most employee pension and welfare plans are covered by a law known as ERISA (Employee Retirement Income Security Act, 1974). ERISA has a number of very specific rules for pension plans. Again, the assistance of an expert in this field is essential.

However, ERISA contains some other, more general obligations. The primary obligation concerns communications to the employees about the plans covered. This communication is called a summary plan description and should be given to employees. Often, the firm you purchase the benefit plan from can give you a summary plan description to distribute to your employees.

FLEXIBLE BENEFITS PLANS

Introduction

Traditionally, employers have offered benefits to employees in a uniform manner. All eligible employees received the same set of benefits in the same way, regardless of their needs, interests, and circumstances. While this method of providing benefits is easier for the employer to administer, it is unproductive: Employees may not want certain benefits their employer offers, and they may want other, more desired benefits that aren't available through their employer.

Recent developments in laws and practices have opened up an entirely new approach to offering benefits to employees. This approach is called "flexible benefits." Other names include: cafeteria benefits plans, flexible spending accounts, and Section 125 plans. Regardless of the name, the essential idea behind flexible benefits plans is to give employees some degree of control over the benefits they receive.

Benefits of Offering Flexible Benefits

Flexible benefits plans require some additional time and administrative effort from the employer. However, other than start-up costs, the overall annual costs of providing benefits need not cost more than what you are currently paying. In fact, in some cases, you may be able to cut expenses.

Here are some additional benefits of flexible benefits plans:

- You attract and retain more satisfied and committed employees because they can obtain the kinds of benefits they want and need.
- You get more productivity out of the benefits dollars you spend. This occurs because you acquire more employee loyalty from the same basic level of expenditure.

Legal or Regulatory Basis

Laws and regulations dealing with how employee compensation can be handled also govern flexible benefits programs. These laws and regulations dictate the way the Internal Revenue Code is applied. For example, there are definite stipulations about how pre-tax dollars must be processed to avoid penalties.

Likewise, there are other benefits laws and regulations, particularly from ERISA, that must be observed and honored.

Get Professional Assistance

The preceding section on legal issues is offered as both a caution and introduction to flexible benefits plans.

Given the complexity and requirements of flexible benefits plans, it is almost impossible for someone not trained in this area to install a program easily and effectively. If you are considering a flexible benefits plan, you should seek professional assistance. Here are several options that may help:

- Contact your current benefits insurer to see what support they can offer you or refer you to.
- Check with any major suppliers or customers to see if they can refer you to assistance.
- Check in the Yellow Pages or on the Internet under benefits, insurance, personnel, human resources, or management consultants.

Simple plans can often be installed using boilerplate documents and procedures that keep costs down. If you do use such documents, seek competent advice to ensure the boilerplate language and procedures are applicable in your state.

In spite of the apparent difficulties and requirements for setting up a flexible benefits plan, the eventual value of such a plan to both you and your employees can easily justify the time and expense.

Types of Flexible Benefits Plans

There are three basic kinds of flexible benefits plan options:

1. **Premium conversion plans**—If you provide health insurance and require some level of employee contribution, it is possible to create a situation that allows the employee to pay for his or her contribution with "pre-tax" dollars. What happens under this plan is that the payment for the contribution is withheld and paid before payroll taxes are computed. In turn, this reduces the taxable compensation level of the employee. Consider an employee making $30,000 per year and paying $2,000 per year for health insurance: Under a premium conversion plan, the employee's W-2 compensation at year end is reported at $28,000 (not $30,000).
2. **Flexible spending accounts (FSAs)**—FSAs operate in essentially the same way a premium conversion plan does. The employee elects to have some portion of his or her compensation withheld on a pre-tax basis, in this case to pay for either medical and/or child care expenses. All of the tax effects of salary reduction described above still apply. With a FSA, the money withheld is credited to an "account" in the employee's name.

   FSAs are different, however, in several ways. First, there are limits and restrictions on how much can be set aside and for what those dollars can be used. For example, there has been a $5,000 limit for dependent care expenses. Second, the employee must claim a reimbursement for expenses incurred, and payment is not automatic. This last point leads to a third matter: FSA dollars in the employee's account exist on a use-or-lose basis. At year's end, dollars not otherwise claimed are lost and cannot be paid to the employee.
3. **Cafeteria plans**—The most flexible—and involved—type of flexible benefits plan is the cafeteria plan. Under this option, each employee is given the choice of how to spend his or her benefit dollars by shopping for the exact kind of benefits desired.

   Say you spend an average total of $5,000 per employee for all of the benefits you offer, which for this example includes health insurance, dental coverage, life insurance, vacation leave, disability insurance, and life insurance. Under a cafeteria plan, each employee would decide how to spend that $5,000. So, to continue the example, the employee might decide to buy a life insurance benefit with a premium value of two times salary while buying medical insurance coverage with a low deductible. The remaining amount could be spent in similar ways for the other benefits.

In order to have a cafeteria plan, you must be able to offer levels of benefits with different prices for each. For life insurance, for example, you could offer plans with premium values set at one time, two times, and three times the employee's annual compensation. There would be different costs for each benefit level.

Planning Issues

There are several important issues to consider in planning and implementing a flexible benefits program:

- Is management committed to providing a flexible benefits plan? A flexible benefits plan requires time, effort, and expense. Further, in the start-up process, employees may be confused and get frustrated. Management must be willing to ride out these waves and be ready to spend the necessary resources to make the program work.
- What level of a flexible benefit should be offered? Relatively speaking, premium options plans and FSAs are easier to install than the cafeteria plans. The first two options can be offered at the same time or singularly. To decide the level of benefit, ask yourself, "What benefits do I wish to have covered?"
- Do you have the administrative capability to manage the plan? You will need to be able to provide an ongoing administration of the flexible benefits plan. This may be done in-house or may be contracted out. Either way, planning for program administration is essential.
- Can you provide appropriate employee communications and training? For both legal and implementation reasons, employees must be educated about the flexible benefits program and be provided ongoing information about it. In some cases, you may obtain training through consulting services. Regardless of the sources, communication is essential.

CONTINUATION OF MEDICAL INSURANCE COVERAGE

Introduction

If you provide some kind of medical insurance benefit to employees, you must attend to Continuation of Medical Insurance Coverage.

Legal Authority

Consolidated Omnibus Budget Reconciliation Act (COBRA).

Are You Covered?

You are covered by this act if you provide a group health or medical plan to your employees, and you had 20 or more employees on a typical day during the prior year.

Action Steps

You must allow an employee, the employee's spouse, and children, if any, the option of continuing their medical insurance coverage after the employee terminates employment with you. This option must be granted to every employee who leaves, with the possible exception of employees fired under narrowly defined and exceptionally egregious circumstances.

1. The employee can be required to pay the full cost of the premium plus a small administrative charge (the total cost to the employee cannot exceed 102 percent of the normal premium charge).
2. You must provide all covered employees and their spouses with a written notice of their continuation rights under COBRA. (See the sample notices on pages 148–176 of Section III. In addition, check with your insurance carrier for appropriate notices you or your carrier must provide to your employees.)

The notice should be provided within 14 days of the employee's separation. Either hand-deliver the notices or send them by certified mail with return receipt.

3. The employee has 60 days to decide what to do. The 60-day period begins with the date he or she receives written notice of the right to elect medical insurance coverage.

4. In general, coverage must be provided for up to 18 months for the employee and up to 36 months for the spouse or other dependents. The length of coverage depends on the qualifying event. Coverage may be extended in other circumstances, so contact your insurance carrier for more details.

Termination

Coverage may be terminated if:

- The employee does not pay the premium in a timely manner.
- The employee becomes covered by another insurance plan, including Medicare. Coverage must continue for employees who are covered by a new plan if they are excluded from certain coverages because of a pre-existing medical condition.

DISTRIBUTION OF COBRA NOTICES

When to Distribute Initial Notices

The initial "Notice of Group Health Continuation Coverage Under COBRA" must be sent to all covered employees and their spouses:

- When a health plan initially becomes subject to COBRA, which means:

  ▲ When the plan is first established. A change in insurance plans does not necessarily constitute the establishment of a new plan. However, we recommend that initial notices be sent to all employees and their spouses when there is a change in plans.
  ▲ When the employer reaches the minimum number of employees for it to become subject to COBRA coverage. To be covered by COBRA, the employer must have employed 20 or more employees on at least 50 percent of its working days during the preceding year. Employers under common control must be aggregated and treated as a single employer for making this determination.

- When new employees are hired. Each time a new employee is hired, a copy of the initial notice must be sent to the employee as well as his or her spouse.
- When an employee adds a spouse. If an employee covered by the health plan marries and adds his or her spouse to the plan, the new spouse must be provided with a COBRA notice.

How to Distribute Initial Notices

1. Initial notices must be provided in a way that accurately informs employees and spouses of their COBRA right. Posting notices on a bulletin board is not sufficient. First-class mailing to employees and their spouses constitutes good faith effort.
2. If the employee and spouse live together at the same address, it's sufficient to send a single mailing, addressed to both the employee and spouse, to the employee's last known address.
3. If the employee and spouse live at different addresses, send the notices in separate first-class mailings.
4. While not specifically endorsed by the government, some employers provide notices to employees by distributing the notices with paychecks. If you do this, you must still send the notice to the spouse by first-class mail.

Inclusion in Summary Plan Description

The initial notice must also be included in the summary plan description of the health insurance plan. This is normally accomplished by including the initial "Notice of Group Health Continuation Coverage Under COBRA."

NOTICES REQUIRED UPON THE OCCURRENCE OF QUALIFYING EVENTS

Covered Employee's Death, Termination, or Reduction in Hours of Employment

- If the qualifying event is the employee's death, termination, or reduction in hours of employment, the employer must notify the *administrator of the health insurance plan of the qualifying event within 30 days* of the occurrence of the event or the date on which insurance coverage is lost, whichever is later. The administrator, in turn, must provide the employee, his or her spouse if covered by the health insurance plan, as well as any dependent children covered by the plan, with a COBRA notice within 14 days from the date the employer notifies the plan administrator. If the employer is also the plan administrator, as is normally the case, the 30-day period is not necessarily lost, although some advisors take the position that if the employer is the plan administrator, the employer must give all covered beneficiaries, including the spouse, a COBRA notice within 14 days of the qualifying event or the occurrence of the qualifying event, whichever is later. In such cases, we recommend that the notice be given by the employer within 14 days of the qualifying event. The administrator or the employer must give a notice of rights and an election form. A sample notice and election form for use when the qualifying event is the termination of the employee or a reduction in the employee's hours is "Notice of Right to Elect COBRA Continuation and/or Conversion of Group Medical Coverage (Termination or Reduction in Hours of Employment)" and "COBRA Continuation Coverage Election Form (Termination or Reduction in Hours of Employment)," which you'll find on pages 148–155 in Section III.
- If the qualifying event is the employee's death, the COBRA notice must be given only to the employee's spouse and dependent children, provided they are covered by the insurance plan. A sample notice and coverage continuation election form for use when the qualifying event is the death of the employee is "Notice of Right to Elect COBRA Continuation Coverage (Death, Divorce, Legal Separation, Medicare Entitlement)," and "COBRA Continuation Coverage Election Form (Death, Divorce, Legal Separation, Medicare Entitlement)," which you'll find on pages 156–162 in Section III.
- When an employee suffers a reduction in hours as a result of disability leave, special procedures must be followed: First, if an employee qualifies for leave under the Family and Medical Leave Act (FMLA) and requests leave for an illness or disability, the employer must determine whether the employee's illness or disability qualifies as a "serious health condition" under the FMLA.
- If the illness or disability qualifies as a "serious health condition" under the FMLA, then the employer may count the leave time against the 12 weeks of annual leave entitlement under the FMLA. This is true even if the employer has a paid sick leave policy or the employee receives disability benefits under a disability insurance policy. If the leave time is counted as FMLA leave, then the employee will not lose his or her health insurance, as the FMLA requires the employer to continue health insurance while an employee is on FMLA leave. However, when the FMLA leave ends, the employer's obligation to continue the health insurance under the FMLA ends, and the employee's COBRA rights commence, unless the employee returns to work. If the employee returns to work, he or she is entitled to continue his or her insurance as an employee. If, however, the employee elects not to return to work, his or her COBRA rights commence. Thus, if the employee does not return to work at the end of the FMLA leave, the employer must notify

the employee, the covered spouse, and all covered dependent children of their COBRA rights within 14 days of the end of the FMLA leave or the date the employer is notified that the employee will not return to work. "Notice of Right to Elect COBRA Continuation and/or Conversion of Group Medical Coverage (Termination or Reduction in Hours of Employment)" and "COBRA Continuation Coverage Election Form (Termination or Reduction in Hours of Employment)." We also recommend that the employer send such an employee the initial "Notice of Group Health Continuation Coverage Under COBRA," which appears on page 147 of Section III, within 14 days of the date the employee notifies the employer he or she wants disability leave.

- If it is determined that the employee requesting the illness or disability leave does not have a "serious health condition" under the FMLA, then the leave may not be counted as FMLA leave and the employer is not obligated to continue health insurance coverage for the employee during the leave. If the employer's policy is to terminate the health insurance of employees at the time they go on sick or disability leave, then the employer must send the employee a notice of COBRA rights and an election form within 14 days of the notice of leave. "Notice of Right to Elect COBRA Continuation and/or Conversion of Group Medical Coverage (Termination or Reduction in Hours of Employment)" and "COBRA Continuation Coverage Election Form (Termination or Reduction in Hours of Employment)" may be used for this purpose. If, on the other hand, the employer's policy is to continue health insurance coverage while employees are on sick or disability leave, the employer need not send a COBRA notice until 14 days after the coverage is lost or the employee notifies the employer he or she is not going to return to work, whichever is later.

- The same special procedures that must be followed when an employee takes sick or disability leave also come into play when an employee takes a leave as a result of a workers' compensation injury. Under the FMLA, an employer can count the time an employee is out on workers' compensation against the 12 weeks of FMLA leave as long as the workers' compensation injury qualifies as a "serious health condition" under the FMLA. This is true even if the employee is receiving workers' compensation benefits under the workers' compensation laws. Thus, in cases where an employee takes time off from work because of a workers' compensation injury that is also a "serious health condition" under the FMLA and the employer's policy is to count the workers' compensation leave against the 12 weeks of FMLA leave, the employer must continue the employee's health insurance during the leave and no COBRA rights are triggered until coverage is lost. If coverage is lost during or at the end of the leave, the employer must give a COBRA notice within 14 days of the date coverage is lost.

If the employer's policy is not to count workers' compensation leave as FMLA leave, then the employer is not obligated to continue health insurance and COBRA rights are triggered by the loss of health insurance coverage at the beginning of the leave. In such cases, the employer must give the COBRA notice within 14 days of notice of the workers' compensation injury. Because of the variables involved in the coordination of FMLA and COBRA rights when employees take sick leave, disability leave, or workers' compensation leave, it is virtually impossible to prepare sample forms for use in such cases. Accordingly, employers should seek advice of legal counsel before preparing their FMLA and COBRA forms to ensure that FMLA and COBRA rights are properly coordinated.

Divorce, Legal Separation, or Cessation of Dependency

- If the qualifying event is a divorce, legal separation, or cessation of dependency, it is the responsibility of either the covered employee, the covered spouse, or one of the covered dependent children to notify the plan administrator or the employer (if the employer is the plan administrator) of the occurrence of the qualifying event. This notice must be given within 60 days of the occurrence of the qualifying event or the date on which coverage would be lost as a result of the event, whichever is later. This notice requirement is satisfied if it is provided by the employee, spouse, or any covered dependent child. It is not necessary for each and every qualified beneficiary to provide notice. If this notice requirement is not met, COBRA coverage does not have to be offered.

■ Once the plan administrator or the employer (if the employer is the plan administrator) receives notice of the divorce, legal separation, or cessation of dependency, the plan administrator or the employer must give the covered spouse and all covered dependent children notice of their COBRA election rights within 14 days of the qualifying event or the loss of coverage, whichever is later. A sample notice and election form to be used in case of divorce or legal separation is "Notice of Right to Elect COBRA Continuation Coverage (Death, Divorce, Legal Separation, Medicare Entitlement)," and "COBRA Continuation Coverage Election Form (Death, Divorce, Legal Separation, Medicare Entitlement)," which appear on pages 156–162 of Section III. A sample notice and election form for use when the qualifying event is the cessation of dependent status is "Notice of Right to Elect COBRA Continuation Coverage (Cessation of Dependent Status of Dependent Child)," and "COBRA Continuation Coverage Election Form (Cessation of Dependent Status of Dependent Child)," which appear on pages 163–168 of Section III.

■ Under COBRA, notice to the covered employee is applicable only to the employee and is not considered notice to anyone else. Notice to the spouse of a covered employee is considered notice to all dependent children living with the spouse at that time. If one or more dependent children do not live with the spouse, then a separate notice must be given to each of the covered dependent children not living with the spouse. Notice to minor covered dependent children must be given to the parent with custody rights or to the legal guardian of the covered child or children.

Employee's Medicare Entitlement

■ If the qualifying event is the covered employee's entitlement to Medicare, the employer must notify the plan administrator of the event within 30 days of the event or the loss of coverage, whichever is later. The plan administrator or the employer (if the employer is also the plan administrator) must, in turn, notify the covered spouse and any covered dependent children of their COBRA election rights within 14 days of receipt of the 30 day notice. If the employer is the plan administrator, we recommend that the spouse and covered dependent children be given this notice within 14 days of the event. A sample notice for this purpose is "Notice of Right to Elect COBRA Continuation Coverage (Death, Divorce, Legal Separation, Medicare Entitlement)."

■ This notice must be given to the covered spouse and any covered children. Notice to the spouse is notice to any covered children living with the spouse. If there are covered children who do not live with the spouse, then each such dependent child must be given a separate notice. Notice to a minor dependent child should be given to the parent with custody or the child's legal guardian.

Bankruptcy of the Employer

An employer's bankruptcy reorganization that causes a loss of health coverage (within one year before or after the bankruptcy filing) for retirees, spouses, and dependent children of retirees and surviving spouses of retirees, creates a special type of qualifying event. Because of the complexities of this type of qualifying event, and the numerous complex legal issues involved, sample notices and election forms are virtually useless and have not been prepared. It is recommended that employers consult an attorney when a bankruptcy reorganization creates a qualifying event as special notices and forms that comply with the bankruptcy code must be developed.

Employee's Retirement

■ The retirement of an employee is considered a termination of employment for COBRA purposes and an employer may treat an employee who retires the same as it would treat any employee who terminates. As in the case of a termination, the employer must notify the administrator of the health insurance plan of the qualifying event within 30 days of the occurrence of the event or the date insurance coverage is lost, whichever is later. The administrator, in turn, must provide the employee and all covered beneficiaries, including the spouse and dependent children, with a COBRA notice within 14 days from the date the employer notifies the plan administrator. If the employer is also the plan administrator, as is normally the case, the 30-day period is not necessarily lost, although some advisors take the position that if the employer is the plan administrator, the employer must provide all covered beneficiaries, including the spouse and dependent children, a COBRA notice within 14 days of the qualifying event or the occurrence of the qualifying event, whichever is later.

■ When the event leading to the loss of insurance coverage is the employee's retirement, notice must be given to the employee, his or her spouse if covered by the health insurance plan, as well as any dependent children covered by the plan. The notice the administrator (or the employer if the employer is the administrator) must give the employee and the qualified beneficiaries, including the spouse and dependent children if covered by the plan, consists of a notice of rights and an election form. As retirement is treated the same as termination of employment, the administrator or the employer may use the same notice and election form used when the qualifying event is termination of employment: "Notice of Right to Elect COBRA Continuation and/or Conversion of Group Medical Coverage (Termination or Reduction in Hours of Employment)," and "COBRA Continuation Coverage Election Form (Termination or Reduction in Hours of Employment)." Alternatively, the employer may use the sample notice and election form "Notice of Right to Elect COBRA Continuation Coverage (Retirement)," and "COBRA Continuation Coverage Election Form (Retirement), which are designed to specifically cover retirees and appear on pages 169–176 of Section III.

Recruitment, Hiring, and Orientation

RECRUITMENT ADS AND NOTICES

Introduction

Recruiting is the process of attracting qualified applicants to fill positions in your company. You should use the most cost-effective recruiting procedures possible, and you should make sure your recruiting process does not violate anti-discrimination laws. A good recruiting program will bring in the kinds of people you need in a timely manner without creating legal problems.

Legal or Regulatory Basis

The full range of equal employment laws apply. In general, the recruitment process should not favor persons of a certain race, sex, age, religion, etc. Recruitment practices that screen out applicants because of their race, sex, age, and/or disability violate equal employment laws and put your business at a legal risk.

Action Steps

Avoid using language in your employment ads and notices that specifies a preference for certain personal characteristics. Examples:

Problematic language	More acceptable
Repairman	Repairperson
Recent trade school graduate	Trade license required
Girl Friday	Administrative assistant
Native-born English speaker essential	Proficiency in English needed

If you are a federal government contractor, you must include a notice that you are an equal opportunity employer (EEO/AA).

Recruiting Tips

See "Candidate Selection Procedures" later in this section.

ACCEPTING EMPLOYMENT APPLICATIONS AND MAINTAINING RECORDS

Introduction

In general, anti-discrimination employment law coverage begins at the job application stage. For this reason, it is prudent to carefully define when and how you will accept applications for employment.

Recommendation

Whatever specific recruiting procedures you establish, you should follow the same procedures for all applicants for the same open position.

- An application for employment will only be accepted when there is a specific open position.
- The applicant must apply for the specific job.

Note: *Employers sometimes wonder whether they should accept resumes instead of written applications. For management or professional positions, you can accept a resume as long as it is for a specific open position. Again, the principle is to be consistent with all applicants for the same position. You can establish different rules for different positions; for example, between professional and non-exempt positions.*

Note: *There is no specific requirement that the application be completed in person. However, there are several considerations here. If you required in-person completion of the application form, you may limit the number of applicants to those who are really interested. And, if you set a time limit on accepting applications, the effect may be the same as requiring an in-person application.*

Recordkeeping Requirements

Employers with more than 100 employees should keep records of the hiring process. In particular, this means that you should keep a record of who applied for and who was hired or promoted for each open position in your company.

Form

A sample hiring record form appears on page 177 of Section III. Make a copy every time there is an open position. Remember, an open position may occur when someone leaves a current position and is replaced or when a new position is created. Also, remember to fill out a form for both outside hires and for promotions. Your recordkeeping forms should include the following information:

- **Position control number**—To make your recordkeeping easy, use a simple numbering sequence to identify each job opening. For example, beginning in January 2004, the first opening could be labeled 04-1, the second opening 04-2, and so on.
- **Application dates**—Indicate the dates during which applications were being accepted, according to the above policy.
- **Applicants**—Simply write in the name of each applicant who meets the qualifications noted above. Remember, this is just a list of all applicants. Record the date on which the application was accepted and note whether the applicant was male or female, and the applicant's ethnic background.
- **Totals**—Add the numbers for each set of categories.
- **Applicant selected**—Indicate who was hired.

APPLICATION FOR EMPLOYMENT

Introduction

A well-developed application and screening procedure is essential for selecting the best people for the jobs you're trying to fill. A good procedure allows you to efficiently compare all qualified applicants. In addition, it helps prevent potential claims of unlawful discrimination in recruitment and hiring.

A well-designed, well-written employment application form is an important tool that helps drive the selection and hiring process. A sample employment application form appears on page 178 of Section III.

Employment Application Guidelines

Make sure the application form only asks for essential, job-related information. In addition, read the following pointers to learn how to address specific issues related to recruiting and hiring employees.

Questions About Age

- **Age of applicant**—Since it is illegal to discriminate against people over 40 years of age, it is not advisable to require information about the person's actual age. Many states ban any discrimination on the basis of age regardless of the person's age.
- **Minors**—Check with your state employment agency for your state's laws about employing teenagers. In general, teenagers 14 to 18 are restricted in when they can work and what types of duties they may perform. Children under 14 are generally prevented from working at any paid job.

  You should ask on the application form if the person is at least 18 years of age. If not, all applicants should be required to produce an age or employment certificate, and/or a work permit.

  The following occupations are particularly hazardous for minors and are detrimental to their health or well-being:

  ▲ Operating an elevator, crane, derrick, hoist, or high lift truck.
  ▲ Work involving riding on a man lift or freight elevator, except a freight elevator operated by an assigned operator.
  ▲ Assisting in the operation of a crane, derrick, or hoist performed by crane hookers, crane chasers, hookers on riggers, rigger helpers, and similar occupations.
  ▲ Operating or assisting in the operation of the following power-driven fixed or portable machines, except those equipped with full automatic feed and ejection: circular saws, band saws, and guillotine shears.
  ▲ Certain roofing operations.

Arrest Record

You should not ask about arrests, as this has been determined to be discriminatory. However, depending on the nature of the work, it may be appropriate to ask about convictions. Check your state laws and regulations.

Disabilities

Questions specifically about disabilities should not be asked.

Strength or Physical Traits

You may inquire whether the person has the ability to do the job, with or without reasonable accommodation.

Notifications

Any false representations on the application for employment are grounds for rejecting the application or for dismissal.

CANDIDATE SELECTION PROCEDURES

Introduction

Hiring a new employee is an important decision for any business. For a company with few employees, it is especially crucial to find just the right addition to your staff to prevent costly turnover and repeated searches.

The selection process also has important public relations significance. When they apply to work for your company, people from the community get an inside look at how you run your business. Discriminatory hiring practices, impolite interviewers, unnecessarily long waits to be interviewed, and neglecting to send follow-up letters can make an unfavorable impression on potential employees who may be your customers some day.

Following consistent recruiting and hiring procedures helps ensure that you bring aboard good, well-qualified employees. This saves you time and money later by reducing performance problems and employee turnover.

Legal or Regulatory Considerations

Every employment law and regulation impacts candidate selection. Whether you're required to or not, it's good business for employers to adhere to these laws and regulations. By doing so, you avoid costly penalties, negative publicity, and lengthy administrative and court processes. Most importantly, you have the opportunity to make full use of the talents present in today's widely diverse workforce.

You should be aware of these important laws and regulations:

- Title VII of the Civil Rights Act prohibits discrimination based on race, sex, color, national origin, and religion.
- The Age Discrimination in Employment Act prohibits discrimination against persons who are 40 years old or older.
- The Americans With Disabilities Act protects qualified individuals with disabilities and individuals who are perceived by employers to be disabled.
- The Immigration Reform and Control Act of 1986 (IRCA) requires that employers verify that all employees are eligible to work in the United States.

State and local laws may augment these laws by prohibiting other types of discrimination.

Action Steps

1. Recruit qualified candidates.

   - Start with a job description. In a sense, job descriptions are pre-qualifying tools. They define jobs and encourage skilled, experienced candidates to apply for them.
   - Advertise the open position. Sources for recruiting include:

     ▲ Your company website
     ▲ Online job boards
     ▲ Newspaper advertisements

▲ Employment agencies and search firms
▲ State unemployment agencies
▲ Colleges and vocational schools
▲ Nonprofit agencies
▲ Employee referrals—tell your employees about the job opening; they may know qualified candidates. (Give employees rewards for referring qualified candidates).

2. Use effective selection devices. The best selection devices allow you to accurately predict how a person will perform on the job. Use devices that can tell you about job performance. For example, rather than just asking if someone can use a certain software program, give him or her a simple problem to work on using that program. The idea is to use short performance tests that require the applicant to demonstrate his or her skills.

   There are several rules for using such tests:

   ■ Give the same test to each applicant who reaches that stage of the selection procedure. For example, if a test is the fourth step in the applicant-screening process, make sure that all applicants who reach this step go through the same test.
   ■ Standardize the testing procedure. Give the same test in the same way to each applicant.
   ■ Explain the procedure and allow a warm-up trial or period.
   ■ If necessary, make a reasonable accommodation for any applicants who request such accommodations. For example, an applicant with a hearing problem may need visual assistance to complete a task.

3. Be consistent. Develop a routine and follow it. There are two important routines to follow. The first is a routine for the entire selection procedure. Plan to use a sequence of selection practices when considering applicants. The sequence may include the following steps:

   ■ Reviewing application forms for minimum qualifications, resulting in a group of the three to five most qualified applicants.
   ■ Interviewing the top candidates.
   ■ Administering a performance test.
   ■ Checking references.

4. Interview candidates. You can't hire employees with job descriptions alone. It's important to meet with candidates to assess their experience and interpersonal skills, and to make sure they're a good fit for your company and vice versa.

   Even though you and a job candidate are getting to know each other, not everything goes in an interview. Some things are unlawful to ask, and doing so may violate a number of U.S. Department of Labor Laws as well as the Americans With Disabilities Act. However, there are plenty of permissible things you can do during an interview to facilitate the process of matching the right person with the right job. Remember that age, gender, race, national origin, and religious affiliations aren't indicators of a person's ability to do a job, so it's a smart idea not to include these topics as part of the interview process. Here are some tips on conducting effective employment interviews:

   ■ Establish a friendly climate:

      ▲ Choose opening questions that are nonthreatening to the candidate.
      ▲ Ask about an item on the resume that you have in common to help establish rapport and relax the candidate.

   ■ State the overall purpose and format of the interview.
   ■ Ask questions and evaluate the candidate's responses to gather information.

Do:

- ▲ Ask questions that assess the candidate's skills, knowledge, and abilities.
- ▲ Ask questions in the same order for all candidates.
- ▲ Ask all the questions of each candidate.
- ▲ Let the candidate answer each question completely before evaluating his or her response.
- ▲ Maintain an objective, yet positive and friendly attitude.
- ▲ Eliminate internal and external distractions and focus on the candidate.
- ▲ Maintain eye contact with the candidate.
- ▲ Listen for central themes, key points, content, and feelings.
- ▲ Let the candidate do most of the talking. He or she should talk about 80 percent of the time, while you talk about 20 percent of the time.
- ▲ Provide information about the position, the company, and anything else that appears relevant.

Don't:

- ▲ Make judgments on the basis of nonverbal cues (e.g., a limp handshake, tapping a foot against a chair).
- ▲ Talk too much.
- ▲ Simply review information on the resume.
- ▲ Overreact to words or ideas that affect you emotionally.
- ▲ Show disagreement or disapproval.

Note: *You may be better off explaining position requirements and expectations* after *the candidate responds to your questions. Otherwise, the candidate may formulate responses to your questions that fit into the criteria you have established for the position.*

- ▪ Tell the candidate what the next steps will be in the selection process.
- ▪ End the interview on a positive note:

    - ▲ Ask the candidate if he or she has any more questions about the position or your company.
    - ▲ Thank the candidate for participating.

Preparing for the Interview

Match the question format to the information you want. Here are three types of questions to ask; each is appropriate at different times during the interview:

- ▪ **Open-ended questions**—Should be used most often. These encourage the candidate to elaborate on his or her feelings or experiences. Use open-ended questions to help determine if the candidate's past behavior suggests success in the job in question.
- ▪ **Hypothetical questions**—Are a variation of open-ended questions. Hypothetical questions are "what if" questions that can help determine how the candidate would handle certain situations. They should always be related to the job requirements.
- ▪ **Closed-ended questions**—Can be answered with "yes," "no," or a short phrase. These should only be asked when you need very specific information.

These general subject areas are valid for questioning:

- ▪ **Knowledge and skills**—All interview questions should be job-related. Candidates should be evaluated on job-related knowledge and skills.
- ▪ **Education and experience**—It's important to probe for and evaluate the content of the candidate's education or work experience. Don't just look at the number of years of education or experience on the candidate's resume.

- **Physical ability**—The employer must be able to demonstrate that a particular physical ability is necessary to perform part or all of the job. Employers are required under law to make "reasonable accommodation" to employ persons with disabilities.
- **"Other" characteristics**—This area includes such things as a valid driver's license, the ability to be bonded, a job-related certification or license, or the ability to work certain hours or under adverse conditions.

Avoid the following types of questions:

- **Leading questions**—These suggest or imply "correct" answers.
- **Stacked questions**—When you ask candidates two or more questions at the same time, they find it difficult to remember the second part of the question by the time they finish answering the first one.
- **Loaded questions**—These have no real right answer. Loaded questions typically ask candidates to choose the lesser of two evils.

Permissible and Illegal Information to Ask for in Interviews

NAME

May ask:

- If the candidate has ever used another name.
- Any additional information regarding an assumed name, changed name, or nickname to verify the applicant's work and education record.

May NOT ask:

- What the candidate's maiden name is.

AGE

May ask:

- If the candidate is 18 years old or older.
- If not, what the candidate's age is.

May NOT ask *until hired:*

- About the ages and birth dates of the candidate's children.

NATIONAL ORIGIN

May ask:

- What language(s) the candidate reads, speaks, or writes (provided foreign language ability is job-related).

May NOT ask:

- About the candidate's nationality, lineage, ancestry, national origin, or place of birth.
- About the candidate's parents' or spouse's place of birth.
- What the candidate's mother or native language is or the language he or she speaks most often.
- How the candidate acquired foreign language ability.

RACE

May ask:

- NOTHING

May NOT ask:

- What the candidate's race or color is.
- Questions about the color of the candidate's skin, eyes, or hair.

CITIZENSHIP

May ask:

- If the candidate is legally authorized to work in the United States.
- If the candidate will require sponsorship for employment visa status (e.g., H-1B visa status) now or in the future.

May NOT ask:

- Which country the candidate is a citizen of.
- If the candidate is a naturalized or native-born citizen.
- What type of visa the candidate has.
- When the candidate became a U.S. citizen.
- The candidate to produce naturalization papers.
- If the candidate's parents or spouse are naturalized or native-born U.S. citizens.
- The date the candidate's parents or spouse acquired citizenship.

GENDER, MARITAL STATUS

May ask:

- For the name and address of a parent or guardian if the candidate is a minor.
- For the name(s) of the candidate's relative(s) already employed by your company.

May state:

- Your company's policy regarding work assignment of employees who are related.

May NOT ask:

- Questions that would indicate the candidate's gender.
- Questions that would indicate the candidate's marital status.
- Number and/or ages of the candidate's children or dependents.
- Questions about a candidate's pregnancy, childbearing, or birth control.
- For names or addresses of relatives, spouse, or children of an adult candidate.

RELIGION

May ask:

- NOTHING

May state:

- Your company's regular work days, hours, and shifts.

Physical Description

May ask:

- What the candidate's height and weight are, but only if the information is commensurate with the specific job requirements.

May NOT ask:

- The candidate to furnish a photograph with the application.
- The candidate, at his or her option, to submit a photograph with the application.
- The candidate to furnish a photograph after the interview but before the job offer.

Ability, Disability

May ask:

- If the candidate can perform the essential functions of the job, either with or without accommodation.

May NOT ask:

- If the candidate has a disability.
- If the candidate has ever been treated for any specific diseases.
- If candidate has, or has ever had, a drug or alcohol problem.

Arrest Record

May ask:

- If the candidate has ever been convicted of a crime. (If he or she has, you may ask for details, but there must be a direct relationship between the job and the offense to use conviction as basis for denial of employment.)

May NOT ask:

- If the candidate has ever been arrested.

Membership in Organizations

May ask:

- About membership in organizations that the candidate considers relevant to his or her ability to perform the job.

May NOT ask:

- The candidate to list all organizations, clubs, societies, and lodges he or she is a member of.

Military Service

May ask:

- Questions about job-related skills the candidate acquired during U.S. military service.
- If the candidate received a dishonorable discharge.

May NOT ask:

- Questions about service in a foreign military.

EDUCATION

May ask:

- About academic, vocational, or professional education schools the candidate attended.

May NOT ask *until hired:*

- Dates of attendance or dates of degrees obtained.

MISCELLANEOUS

You should not ask:

- Questions about financial credit.
- Questions about financial status.
- Questions about union membership.

Examples of Effective Interview Questions

- Tell me what you already know about the position.
- Describe your previous (or current) position. What percentage of your time was or is spent on each function?
- What experiences have you had that you would like to increase in your next position?
- Which of your experiences would you like to decrease?
- How do you prepare yourself for the work that you do?
- Describe a scenario of a situation this candidate will encounter and ask the candidate how he or she would handle the situation. Or, ask the candidate to describe his or her experience in a particular situation, and to provide detailed information about how he or she actually handled it.

An interview offers the opportunity to evaluate a candidate's verbal communication skills, mental alertness, personal appearance, and body language. If time permits, and *only if* you intend to do this for each interviewee, ask for a follow-up letter that demonstrates written communication skills. For example, "Candidate, I would like you to write a one-page description of why this position interests you, and which of your key attributes will contribute to your success in this position."

After the interview, record your impressions of the applicant's qualifications on a separate piece of paper. Rate the candidate according to the qualifications for the position. At this point, do not compare candidates to each other.

Note: *It is not advisable to record information and impressions of an individual on a job application, resume, or other document completed by the applicant. Information, opinions, and impressions can be misconstrued and interpreted to mean many things other than what you meant by them. Those notes may be used against you should litigation arise.*

Hire the Most Qualified Person

Once you have interviewed enough candidates to make a good decision (maybe three or four for a clerical position or as many as 10 for a management position), review each candidate's qualifications. Decide which candidate(s) to bring back for second interviews or to meet with other managers to see how the applicants relate to other people.

Complete reference checks on the top candidate(s). Verify all information provided by the candidate, including education and past employment dates and salaries. Try to talk directly with the candidate's past supervisors. (See "Checking References" later in this section.)

Before making the offer, determine the salary and the starting date. Once the candidate has accepted your job offer, notify the other candidates that a decision has been made either by mail or phone. Thank them for the time they spent with you. (See "Sending Offer of Employment Letter, Rejection Notification" later in this section.)

Monitor Selection Rates

If you find that your procedure is screening out women applicants, for example, you may have a problem. Examine your selection procedures to see if any aspects of it may be creating discrimination problems. If so, correct or revise the procedures.

You can use both the sample hiring record form and the sample applicant data record found on pages 177 and 181 of Section III to monitor selection rates.

CHECKING REFERENCES

Introduction

A simple reference checking procedure can help identify potential problems before hiring an employee. While not directly required by law, reference checking is strongly recommended. States are becoming more active in regulating reference checks, especially on the issue of requiring an applicant's written authorization to check references. (Check with your state's attorney general's office on this matter.)

Types of Reference Checks

You can use several methods to check job applicants' references. Select the method that best suits the job for which you are considering the applicant. For example, if you're hiring someone to handle money or deal extensively with customers, you'll probably need to check the applicant's references more extensively than you would if you were filling a day laborer position.

- **Personal friends, family, or acquaintances**—Generally, references from these people are not helpful and should be avoided. They're often biased due to the individual's relationship with the applicant.
- **Education or training**—Use either the employment/education reference verification form on page 182 of Section III or require the applicant to have an official transcript sent to you from the last institution he or she attended.
- **Prior employers**—This is an essential reference check. Ask for the name of the applicant's direct supervisor.
- **Credit checks**—This is a good reference to check if the employee will be handling cash or money. Credit bureaus may release credit reports to employers for the purpose of evaluating candidates. However, if you use a credit report and do not hire the person, you must notify the person that you considered their credit report as part of your decision not to hire. This must be done whether or not the information in the report had any bearing on your decision or not. Refer to your state laws and regulations regarding credit reporting laws and notification.
- **Driving records**—You may be able to obtain an applicant's driving record through your state's department of motor vehicles. Contact that department for information. Checking driving records is especially important if the employee will be driving any of your company vehicles.

- **Criminal records of conviction**—This may be an appropriate reference to check if the applicant would work in a sensitive or unsupervised position. Contact your state department of corrections or public safety for details.
- **FBI fingerprinting**—This may be appropriate under certain situations for positions that may require fingerprinting. Contact your local FBI office for details.

How Long Ago?

The more distant the reference, the less its current value. In general, employment references more than five years old may have limited value. However, education records can and should be verified. Other records may still have value, even if they are dated. For example, driving records can be important for positions that involve driving. Such records, even though old, should probably be examined.

When Do You Check References?

Because reference checking requires time and effort, it's best to wait to until applicants have passed initial hiring screening and appear to be serious potential job candidates before checking their references.

However, you can prepare for the process by requiring all applicants to complete an employment/ education reference verification form (which appears on page 182 of Section III) when they fill out employment application forms.

Having applicants fill out and sign an employment reference verification form at the very beginning lets them know that their references will be checked. If an applicant is unsure about what his or her references might say, the person may drop out of future consideration. This saves you time by narrowing the pool of serious candidates.

Be Consistent

Once you know which references you will check, be sure to follow that same procedure with every applicant.

What About Giving References?

Other employers may contact you for references on people who used to work for you. Keep the following rules in mind:

- You should only verify the position's title and the dates the person worked for you. Supplying any other information is risky and may make you liable in a defamation suit.
- Wait for the caller to ask for information; do not volunteer any information without being first asked for it.
- Make sure the information you supply is accurate; check your records if necessary.
- Do not offer comments or opinions about the person's character.
- Some states now have laws that protect employers when they provide reference check information, as long as the information was accurate and/or made in good faith. Check with your state's attorney general's office for details.
- You may want to have an agreed-upon reference statement for an employee who is leaving the company. Make sure you and the employee agree on what will be told to prospective employers, and that you agree on the reason for the termination. The reference statement should be a written document that the employee signs and dates. Make sure it is the only reference used.

SENDING OFFER OF EMPLOYMENT LETTER, REJECTION NOTIFICATION

Offer of Employment Letter

An offer of employment letter is a simple yet important way to make sure the employment relationship gets off on the right foot. By using such a letter, you can prevent misunderstandings about the position title, compensation, start date, and other details. This correspondence should be printed on the firm's letterhead and should be addressed to the new employee. A sample offer of employment letter appears on page 183 of Section III. Be sure also to use the offer of employment letter attachment, which appears on page 184 of Section III.

Rejection Letter

You are not required to notify applicants that they have been rejected for employment. However, here are two good reasons for notifying them:

- It is a considerate gesture that may help build or retain goodwill in the name of your business.
- It may save you time from answering calls from applicants.

Action Steps

1. Create a set of pre-printed small post cards to be mailed to the applicants.
2. The message should explain that while the applicant had many strong qualities, you picked another applicant with qualifications that matched the requirements for the job more completely. Indicate that you appreciate the applicant's interest in working for you and that you wish the applicant the best in his or her job search.
3. If you plan to keep the applicant's resume and/or application on file for a certain period of time (some builders do this in case a more suitable position opens up), say so in the rejection letter or post card.

VERIFYING EMPLOYMENT STATUS: THE I-9 PROCEDURE

Introduction

You are required to verify that the employee who will be working for you is legally eligible to do so. As an employer, you must verify that the new employee is either a citizen of the United States or is authorized to be employed here. If you hire an illegal alien, there can be stiff fines imposed on your business.

Legal or Regulatory Basis

The Immigration Reform and Control Act of 1986 applies to all employers.

Action Steps

You must complete an I-9 form of employment eligibility verification. A copy of the form appears on pages 185–187 of Section III.

1. The employee must complete Section 1 and sign the form. The employee must provide you with suitable verification documents.
2. You are to review the documents provided, noting any relevant details about the document(s) in the space(s) provided. You will sign Section 2, which indicates that the documents appear to be authentic.

Deadline

This process must be completed within three working days after hire. However, since Section 1 must be completed by any new employee on the first day of employment, it makes sense to complete the entire process on that same day.

Rehires

For employees who are laid off or quit, but are rehired within three years, you may simply verify that the supplied documents are still correct. Rehires after more than three years must complete a new I-9 form.

Lapsed Verification

Employees working under a temporary employment authorization (foreign workers with limitations on the dates they are authorized to work in the United States) must be reverified when their authorization expires, or be terminated. This can be done in Section 3 of the I-9 form on file or on a new form.

NEW EMPLOYEE ORIENTATION

Introduction

An employee orientation program helps you form a positive working relationship with each new employee. By establishing expectations and communications at the outset, you increase your chances of maintaining a productive and constructive employee relationship over the long term.

Legal or Regulatory Considerations

The law mandates only a few orientation and training requirements, such as safety and hazardous materials communications, drug-free workplaces, and prevention of sexual harassment. Some of the noted requirements are a function of employer eligibility (e.g., communications about drug-free policies for employers with federal contracts) or state law (e.g., sexual harassment policies).

Action Steps

1. Establish and communicate employment expectations, job responsibilities, duties, and performance standards.
2. Help the employee to become productive as quickly as possible while becoming a member of the work group or team.
3. Build the basis of a good working relationship.
4. Communicate safety policies and procedures so the employee becomes familiar with them and understands how the measures protect him or her.

Orientation Program Basics

An orientation program explains the way your company does things. Be sure to communicate procedures for the following functions to new employees:

- Job duties and expected performance standards
- Performance evaluations
- Business mission and goals
- Daily policies and procedures (e.g., calling in sick)
- Company products and services

- Dealing with customers
- Working with others on the team or work group
- How to work with the supervisor effectively

It is unreasonable to expect that you can cover all these items or that the employee can remember all this information with a first-day orientation. An effective orientation program should take place during the employee's first few weeks on the job. Most of the orientation process should be handled by the employee's direct supervisor, whom you have trained.

Use the new employee orientation procedures checklist on page 188 of Section III to make sure you cover all necessary and important matters with the new employee. Put the new employee's name at the top of the sheet, check off items as they are completed, and retain the sheet in the employee's personnel folder.

Managing Employee Safety

MAINTAINING A SAFE WORKPLACE

Introduction

In general, every employer has an obligation to provide a safe and healthy workplace for its employees. Not only does this comply with the law, but a safe and healthy workplace benefits your business in other ways:

- You avoid and minimize the direct expenses of employee injuries and the indirect costs of insurance.
- You keep your employees productive on the job.
- You demonstrate that you care about your employees' safety and well-being, which boosts their morale and can increase their feelings of goodwill toward your company.

Legal or Regulatory Basis

The Occupational Safety and Health Act extends to virtually every employer in the country; self-employed persons are exempt. This law is administered by the Occupational Safety and Health Administration (OSHA). Some states have their own OSHA program and standards, but to have such a program, their standards and practices have to be at least as effective as OSHA. Check your state laws and regulations for requirements.

Employee Rights

Under the Occupational Safety and Health Act, employees can refuse to work in environments they consider dangerous and/or life threatening. They can call OSHA to report hazards, which may result in an inspection. They can pursue these rights without fear of retaliation or discrimination and can ask for an investigation if they believe this right has been abridged.

What OSHA Can Do

Under the law, OSHA may inspect your worksite to look for safety hazards, may set standards for safe practices, and may issue fines and penalties.

An OSHA Inspection

An OSHA inspector will arrive at your worksite for an inspection without advance notice. If this happens, ask to see the inspector's credentials and verify their authenticity. While the inspector has the right to enter the worksite at that time, you may request that the inspector wait for a few minutes until a senior officer from your company can arrive. You can request a warrant for entry, but this very important decision should be made with legal assistance.

The inspector will brief you on the nature and focus of the inspection. This may include talking to employees, which can be done privately. If the inspector has been summoned by an employee, that employee may remain anonymous. The inspector may take pictures, collect samples, and videotape evidence.

The onsite company representative may—and should—accompany the inspector during the site inspection. You should take notes about what happened and what was said by both the inspector and you; also take pictures of the same evidence the inspector photographs.

Finally, after the inspection, the inspector will confer with you about any alleged hazards found.

What You Must Do Under the Law

If you have 10 or more employees at any time during the year, you must maintain records of occupational injuries and illnesses at each business location. Use the following recordkeeping forms:

- **OSHA Form 301—Injury and Illness Incident Report.** This is one of the first forms you must fill out when a recordable work-related injury or illness has occurred. Together with the Log of Work-Related Injuries and Illnesses and accompanying Summary, these forms help the employer and OSHA develop a picture of the extent and severity of work-related incidents. Within seven calendar days after you receive information that a recordable work-related injury or illness has occurred, you must fill out this form or an equivalent. Some state workers' compensation, insurance, or other reports may be acceptable substitutes.
- **OSHA Form 300—Log of Work-Related Injuries and Illnesses.** The Log of Work-Related Injuries and Illnesses is used to classify work-related injuries and illnesses and to note the extent and severity of each case.
- **OSHA Form 300A—Summary of Work-Related Injuries and Illnesses.** The Summary shows the total number of injuries for the year in each category. At the end of the year, post the Summary in a visible location so that your employees are aware of the injuries and illnesses occurring in their workplace.

Note: *Please see pages 189–200 in Section III for copies of OSHA forms 301, 300, and 300A, and for detailed instructions and examples for filling them out.*

Refer to your state and local laws and regulations. Most state laws require filing of OSHA records with a state agency, from the employer's first report of injury.

If an on-the-job accident occurs that causes the death of at least one employee and/or the hospitalization of three or more workers, you are required to report the accident in detail to the nearest OSHA Area Office or state agency.

You are required to post a notice or poster announcing safety protections in a conspicuous place. Go to www.osha.gov to view and order notices and posters.

What Injuries Must You Note?

An occupational injury or illness must be recorded in the OSHA Form 300—Log of Work-Related Injuries and Illnesses if it involves:

- One or more lost work days to the employee
- Restriction of work or motion by the employee
- Loss of consciousness
- Medical treatment beyond first aid
- Transferring the employee to another job
- Death

DEVELOPING AN EMPLOYEE SAFETY PROGRAM

Introduction

Use your site superintendents' input to help you develop an employee safety program. After all, they are instrumental in implementing the program and seeing to it that field employees follow established safety procedures. Here are some additional pointers for developing a safety program for your company.

Action Steps

1. Reduce workplace risks and dangers.

   - Make sure your supervisors are properly prepared for managing a safe workplace and handling any accidents or emergencies. They should be trained in all areas of hazard identification and control, such as first aid, risk management, and safety orientation and training.
   - Have your site supervisor conduct regular jobsite inspections using a safety checklist or other form you provide. These inspections should be done regularly and routinely. Have the supervisor send the completed checklist to you for review. A sample jobsite safety audit checklist appears on pages 201–203 of Section III.
   - Supervisors should thoroughly investigate and report accidents to a senior officer of the company at the worksite to help avoid those accidents in the future
   - See if your insurer provides consulting assistance or other resources for workplace safety.

2. Be prepared.

   - **First aid**—Make sure there is a well-stocked first aid kit on all worksites. Have at least one worker certified in first aid on site at all times.
   - **Medical emergencies**—Post phone numbers for emergency services at a convenient location on the jobsite. Have at least one worker trained in dealing with medical emergencies on site during all business hours.

3. Train your employees to be safe.

   - Provide training in the proper use of all safety equipment. Be open to employees' suggestions about additional safety equipment for use on worksites.
   - **First aid training**—Contact your local chapter of the American Red Cross or another qualified group to train your supervisors and/or employees in how to provide first aid.
   - **Periodic safety talks**—Have supervisors regularly provide short (e.g., 15 minute weekly) training sessions on selected topics. The training can take place at the start of the shift, after a break, after lunch, or whenever it is appropriate. The purpose is to remind the workers of specific, relevant safety procedures and to build awareness of the importance of safety. These meetings can also be used to solicit employee ideas for safety improvements and to discuss new hazardous materials that employees may encounter on site. Topics may include fall protection, lifting and handling objects, chemical hazards, eye protection, working around electricity, safe use of various tools or equipment, heat exhaustion, scaffolding, and ventilation.

4. Communicate your safety policy to employees.

   - Display the safety protection poster in a conspicuous place. (See "Required Posters" later in this section.)
   - Post completed OSHA Form 300A in a visible location and discuss it with employees.

- Include your safety policy and procedures in your employee handbook. New employees should be specifically briefed about your company's safety policies and procedures as part of the employee orientation.
- Establish and communicate consequences for lack of employee adherence to safety standards and lack of or misuse of safety equipment. Discipline may include termination of employment.
- Make sure your employees receive proper training about any hazardous materials they may be exposed to. Update employees as each new hazardous material is introduced into the worksite. (See "Hazard Communications" later in this section.) Train your employees on how to deal with hazards on multi-employer worksites.

5. Establish a risk management program.

- Have regular meetings with employees and supervisors to review safety matters. Use these meetings for planning safety toolbox talks, for promoting safety in the business, and for studying accident investigation.
- Set specific safety goals, such as reducing incidents of injuries to zero within a certain time frame. The more specific the goal, the better. Impress on your employees that you are committed to this goal; they should know that this is a serious objective you intend to accomplish.
- Use a safety incentive program that rewards employees for safe job performance. For example, you could create a program that provides a reward (e.g., a catered lunch or a special bonus of $25 per employee) if everyone is observed following all of the safety policies for a specific time period.
- Encourage employee suggestions and initiatives; ask your employees for their ideas and suggestions for making the workplace safer. You could create an incentive program that pays a bonus (e.g., $50) for useful suggestions that lead to safety improvements.

6. Control costs.

- Once you have developed a safety management program, put it in writing and send a copy to your insurance broker or carrier. In some states, you can receive a discount on your insurance for having a written safety management program that prevents accidents.

For More Information

The National Association of Home Builders' Labor, Safety & Health Services department can provide you with additional information to help you run a safer, OSHA-compliant business. Call 800-368-5242 or check out the department's resources online at www.nahb.org.

You'll also find the following books useful:

- Published by NAHB, *The Construction Safety Program Manual: A Guide for Home Builders and Contractors* contains tools and information to help you set up a safety program for your company.
- *NAHB-OSHA Jobsite Safety Handbook* (available in English and Spanish editions)
- *Toolbox Safety Talks,* English-Spanish edition

All three are available from BuilderBooks. Call 800-223-2665 or visit www.builderbooks.com to order them online.

In addition, check out *Selected Construction Regulations for the Building Industry,* available online through OSHA at www.osha.gov.

OSHA offers a no-charge, non-punitive consultation service to help employers identify and solve safety problems. Call NAHB at 800-368-5242 to find out how to contact your state or local agency.

HAZARD COMMUNICATIONS

Introduction

If you use hazardous materials in the workplace, you are required to provide notification and training about the materials to your employees. This is required by OSHA's Hazard Communication Standard (Hazcom).

Are You Covered?

You are covered if you use hazardous materials in your work.

What Are Hazardous Materials?

Hazardous materials are any substances or compounds that can produce an adverse effect on the safety or health of a human being. This can include carcinogens, irritants, corrosives, poisons, flammables, etc.

Action Steps

You have several obligations under the Hazcom standard:

1. You must make sure hazardous materials are labeled in your workplace. Except for gasoline, this usually is done by the supplier of the chemicals.
2. Collect and store on the worksite the Material Safety Data Sheets provided with hazardous materials.
3. You are required to prepare a hazard communication program for employees. This should be a written plan that includes information about:

   ■ The Hazcom standard
   ■ Where to find the Material Safety Data Sheets
   ■ A list of any known hazardous materials in the workplace
   ■ How to follow safe procedures when working with hazardous materials
   ■ Protective measures and available equipment

An outline for a sample plan for hazard communications appears on page 204 of Section III.

A sample communication on the Hazcom standard is included as part of the sample employee handbook in Section II.

A 1994 law change requires *written certification* that the employer has performed a workplace assessment and has identified potential hazards and necessary personal protection equipment.

Employee Communications

REQUIRED POSTERS

Introduction

You are required to post certain notices for employee inspection under the various laws affecting employment. These posters are available from the various agencies, at either the federal or state level.

What Information to Post

You should display notices regarding the following issues. The posters should be located in a place that is accessible and convenient to employees:

Issue	Law
Employment discrimination	Civil Rights Act Age Discrimination in Employment
Minimum wage	Fair Labor Standards Act
Lie detectors	Polygraph Protection Act
Job safety	Occupational Safety and Health Act
Disabilities	Americans With Disabilities Act
Family leave	Family and Medical Leave Act

Check to see if state and local laws and regulations require you to display additional posters and/or information.

GRIEVANCE PROBLEM-SOLVING PROCEDURE

Introduction

None of the employment laws require an employer to institute some kind of internal problem identification and grievance resolution procedure. However, there are two important reasons why it's important to have such a system for employee communications:

1. Being open and responsive to employee concerns and problems gives you the chance to address them. Without a way to listen to employees and learn about their concerns, you cannot respond. Listening and addressing employee concerns builds a loyal and committed workforce.
2. A problem-solving procedure is a critical element in confronting sexual harassment, discrimination, and similarly serious problems, and can be indispensable in avoiding liability if someone brings a charge against your business on this matter.

What Should an Effective Procedure Contain?

An effective problem-solving communication procedure should include the following elements:

- A way for employees to raise problems and concerns privately and confidentially; this should include an assurance that the employee will be not be penalized or harmed for bringing forth concerns in a good-faith manner.
- A way to discuss problems or concerns with someone other than the employee's immediate supervisor, if necessary.
- A rapid acknowledgment of the employee's problem and a promise of a quick and well-considered response.
- A thorough investigation of all employee problems.
- Appropriate action to correct the problem where that is indicated.
- Communication with the employee indicating how the matter is being handled; this may include some kind of appeal or further review procedure.

Suggested Procedure

We suggest instituting procedures to support the following grievance problem-solving procedure policy statements.

- Employees are encouraged to bring up any problems or concerns they are experiencing. They will not be penalized for raising problems or concerns in a good-faith effort. While it may not be possible to guarantee their confidentiality and still address the problem, every effort will be made to protect employee privacy in this process.
- Your company will make every effort to address the problem quickly and thoroughly. Where possible and appropriate, we will take necessary action to resolve the problem fairly. The employee will be informed of the results and reasons for them.
- If you are experiencing some kind of problem relating to your work, discuss the problem with your immediate supervisor if possible. If that is not possible, you may ask to discuss the problem with some senior official of the company.
- If you do not receive a response to your problem from your immediate supervisor within a reasonable time or at all, you should bring this matter to the attention of the company president.
- If you are not satisfied with the response to your problem, you may ask that the response be reviewed and reconsidered by a senior official of the business, including the president.

A sample policy on this matter is included in the employee handbook in Section II.

PERFORMANCE APPRAISAL SYSTEMS

Introduction

The purpose of a performance appraisal is to meet privately with an employee to talk about his or her continued success and growth and that of your company. Although day-to-day performance feedback is important, it should not take the place of a formal, written appraisal conducted on a regular basis.

A written performance appraisal documents the discussion of past performance and future expectations between an employee and a supervisor. The employee, the supervisor, and the company benefit from this discussion. A performance appraisal system consists of the following:

- Ongoing, frequent feedback on performance
- Documentation of performance evaluation
- Documentation of employee warnings, disciplinary actions, or reprimands
- Periodic review
- Formal, written appraisal

Many companies do written performance appraisals once a year, usually on the anniversary of the employee's hire date or at the end of the fiscal year. Some do them quarterly or semiannually. It's best to start out doing appraisals annually so you don't become overwhelmed with filling out forms.

Legal or Regulatory Considerations

The Equal Employment Opportunity Commission (EEOC) and other federal enforcement agencies make it clear that performance appraisals must be job-related and nondiscriminatory.

Action Steps

1. Keep short, factual notes of incidents that you observed and discussed with the employee during the year. Discuss each note with the employee at the time it is written and have the employee sign and date it. These notes become the starting point for the written appraisal and ensure that nothing will surprise the employee at the end of the year. The notes also protect the employer; with his or her signature on each document, the employee cannot later claim that he or she never saw it.
2. Take the time to complete the written appraisal carefully. It is a formal document that is part of the employee's records. The evaluation should be fact-driven and based on your own experience and observations, not on gossip or second-hand observations. Write only factual statements on the evaluation. Avoid recording opinions and unsubstantiated rumors. Don't include terms like "never," "always," "wonderful," "exceptional," "outstanding," "excellent," "perfect," "gifted," "appears," or "in my opinion" in the appraisal. A sample performance appraisal form appears on pages 206–208 of Section III.
3. Set up the performance appraisal meeting with the employee a few days in advance. Give the employee the opportunity to prepare for the meeting by reviewing his or her own performance since the last review and considering goals for the next year. A sample performance review preparation form appears on page 209 of Section III.

Appraisal Meeting Guidelines

First discuss the completed performance appraisal form. Your assessment should be descriptive and provide specific performance examples. Ask the employee how he or she could improve performance. Then suggest additional techniques or approaches that may help to improve performance.

Ask for the employee's response and comments. Telling the employee what to do or not to do will have less impact than helping him or her develop an action plan for the future. Complete agreement is not necessary as long as each person fully understands the other person's views. Emphasize the future instead of belaboring past results and events.

The written performance appraisal should not be changed during the meeting unless the employee presents information you were not aware of. Even then, your written evaluation should not be changed, but additional comments may be added.

Ask the employee if he or she wants to add comments to the evaluation, and have the employee sign the performance appraisal form. You want the form signed to document the fact that the employee received the appraisal and that you discussed it with him or her. You should include the following statement above the employee signature:

"I acknowledge receipt of the evaluation and it has been discussed with me. My signature does not imply that I agree or disagree with the content."

Make a copy of the appraisal for the employee and keep the original in the employee's personnel file.

At the meeting's conclusion, summarize what was discussed and what you and the employee agreed upon. Review plans for the future and set specific goals regarding what each of you will do. Employees should sign all documentation relating to future plans and goal setting. When needed, set follow-up dates to meet and review progress. Praise the employee for work well done and end the meeting on a positive note of encouragement.

Discussing Unsatisfactory Performance

A performance appraisal discussion with an unsatisfactory performer can be a very difficult meeting. Since you will be discussing the employee's poor performance, he or she may become hostile, defensive, or argumentative. The employee may challenge your assessment of a particular rating or the overall appraisal. You can overcome these problems if you are thoroughly prepared for the discussion and:

- You have had previous discussions throughout the year with the employee about his or her inadequate performance. Nothing should come as a surprise during the performance appraisal meeting. All performance issues discussed with the employee throughout the year should be documented and signed by the employee.
- You can provide accurate information to justify the ratings.
- You are open to the employee's concerns and respond empathetically.

It is important to focus the discussion on specific performance. A performance appraisal is an evaluation of the employee's performance, not a judgment of his or her value as a person.

Be prepared to supply additional data, if necessary, to ensure that the information discussed is accurate and thorough. Focusing on specific performance issues and reviewing previous discussions will help to create a problem-solving climate in this appraisal meeting. When the employee's overall performance is less than satisfactory, he or she must understand that performance must improve. Therefore, you should see to it that the goals set meet the position's minimum requirements, and explain that the employee is expected to meet them in a specific amount of time.

The employee will be more likely to commit to improving performance if:

- You "level" with him or her about the consequences of continued unsatisfactory performance, including suspension or termination. You must be prepared to follow through with these consequences if the goals are not met.
- You listen openly and respond with empathy to the employee's explanations and feelings.
- You involve the employee in a discussion of causes and possible solutions for those areas rated less than satisfactory.

Your final objective is to agree on actions that will help the employee achieve minimum acceptable standards of performance within a reasonable time frame. If you can't come to an agreement, you must still outline what actions the employee must take to meet minimum performance standards.

DEALING WITH EMPLOYEE PERFORMANCE PROBLEMS

Introduction

There may be times when an employee's job performance does not meet your expectations or requirements. Employee performance problems indicate that you are not receiving the level of productivity for which you are paying the employee. Further, you may be spending additional time and effort in trying to correct the problems. Other employees may also be adversely impacted and may resent poor performers receiving the same compensation they do.

You should correct performance problems decisively and consistently without creating additional problems. If you simply terminate an employee who has a performance problem without first investigating and/or addressing the problem, you may create morale problems with other employees. This could occur, for example, if you asked the employee to do a new task without proper training or support, and then fired the employee at the first sign of a problem and without any discussion. Any perceived inconsistencies in disciplinary actions may create questions about unlawful discrimination in violation of federal, state, and local laws.

Legal or Regulatory Considerations

There is no law that specifically deals with employee performance problems. However, the various civil rights laws that protect employees from unfair treatment based on race, color, gender, age, religion, national origin, handicap, etc., do apply here.

In general, make sure that any actions you take for performance problems are applied consistently for all employees and are based on sound business reasons. Avoid the appearance of basing your actions against the employee based on a protected attribute, such as race, age, gender, etc.

Action Steps

1. Identify and correct the cause of the problem. Meet with the employee to discuss the problem and find out the employee's perspective. Be open to the possibility that you may be responsible in part for the performance problem or that the cause of the problem may be something other than poor employee motivation or carelessness. The employee may not have the proper training, equipment, or safety equipment to adequately perform the job requirements.
   For example, consider whether any of the following factors may be causing poor performance:

   - Unclear performance expectations
   - Uncertainty about what duties the employee is expected to do
   - Lack of adequate skills due to inadequate training or coaching
   - Poorly designed work procedures
   - Inadequate tools or equipment
   - Insufficient feedback and/or compensation
   - Need for reasonable accommodation due to a disability
   - Unsafe conditions

2. Be ready to serve as a coach to help provide more training. You may also need to change how you manage by modifying your expectations, helping reorganize tasks, providing more or better information, and/or by obtaining better equipment.

3. Treat the employee fairly and prevent morale problems. Tell employees up front what kinds of actions or behaviors will not be tolerated. Explain the details in the employee handbook and during new employee orientation.

4. Adopt and follow a progressive disciplinary procedure. A procedure means that you generally follow certain steps when responding to employee problems. This builds in a sense of consistency in how you treat employees. It also assures employees that they will be treated fairly and in a non-discriminatory manner.

5. Make sure that the performance problems are job-related. Criticism of trivial or non-job related actions only invite further difficulties. You cannot discipline an employee for non job-related or off-job problems or incidents that do not impact employee performance on the job.

Prepare the Way for Disciplinary Action, Including Termination

Be sure to investigate the problem before taking action. You may need to:

- Interview other employees.
- Collect written statements, dated and signed.
- Take into custody any relevant items, such as stolen property or records.
- Remove and/or suspend the employee from work during the investigative period.
- Document the entire investigative process, including actions taken, the date of such actions, and information learned. Be factual in your documentation. Avoid expressing opinions of guilt/innocence during the investigative process and in the documentation.

You should also give the employee the opportunity to tell his or her side of the story. Do so at the start of the investigative process and again throughout the process as necessary. Be sure to make notes.

If the employee is in a protected category (e.g., minority, female, over 40, disabled, etc.), consider how you have handled other employees in the past who have done similar things and take the same action. If you find that you are treating this employee differently, reconsider what you are doing and follow your set policies and procedures.

Avoid Employment Discrimination Liability

If necessary, consult with a human resources management professional or legal counsel prior to acting. Here are some additional tips:

- Document everything relevant. Make sure you keep the records and documents together and safely filed. Keep a written record, such as short, handwritten notes, of all performance-related discussions you have with the employee.
- Never summarily discharge an employee. Firing an employee on the spot may be gratifying at the time but it can lead to problems. Regardless of what you catch the employee doing, it is better to put the employee on suspension pending an investigation, with or without pay. You can always terminate an employee at a later date. Make sure you document the suspension, all investigations, and the final decision.
- Unless circumstances are unusual or severe, you should follow the disciplinary procedures you have laid out in your personnel policies and procedures handbook. The disciplinary procedure followed should include the following steps:

  - The employee should be notified that there is a job-related performance problem.
  - The employee should be told what the expected performance standard is, why he or she is not meeting that standard, and the allotted time in which to improve his her performance.
  - The employee should also be told that failure to improve the performance by the time set may lead to further penalties, up to and including termination.

A sample notice of unsatisfactory performance appears on page 210 of Section III.

Alternative Approach

The method described above is known as a "progressive disciplinary" policy. The rationale for this approach is that each additional instance of a performance problem is met by an increasing penalty.

A recently advocated alternative approach is called "discipline without punishment." Proponents claim this approach has better results than the traditional progressive disciplinary approach.

In general, the approach works like this:

1. After the first instance of the problem, the supervisor meets with the employee to note the problem and remind the employee of his/her personal responsibility to meet reasonable performance standards. The supervisor tries to have the employee agree to solve the problem. No threat is made but a note of the discussion and what was decided is kept in the employee's personnel file.
2. If the problem occurs again, another discussion takes place covering the same matter. The supervisor explains why the standard is important and seeks employee agreement to solve the problem. The two develop an action plan for solving the problem. The supervisor writes a note summarizing the discussion and what is decided, and the note is kept in the employee's personnel file.
3. If there is no performance improvement within two weeks of the discussion, the employee is told to choose between committing to solve the problem or quitting, and that any further instances of the problem will lead to termination. Document this meeting.

TERMINATING EMPLOYEES

Introduction

Regardless of who decides to end the employment relationship, employee termination can produce current or future problems. A well-planned and -executed procedure for terminating employees gives your operations and management legal and professional support. Employee termination gives you an opportunity to discover why an employee is leaving—and thereby learn how you can improve your business and retain employees in the future.

Legal Issues

There are several legal issues to keep in mind when terminating an employee:

- **Proper final wage payment**—Your state law governs payment for wages and/or leave (such as the timing of the last paycheck and unused vacation or sick leave), and when final payment of compensation due the employee needs to be made. (See the information on "Compensation Program and Administration" earlier in this section.) Check with your state authority on any specific laws and regulations on this matter.

   A good business practice is to have all wages and other compensation payments due to the employee ready and available at the time the employee is terminated

- **Medical insurance continuation**—If you offer medical insurance, you'll generally be required to offer the employee the opportunity to continue his/her coverage. (See the information on "Employee Welfare Plans" earlier in this section for your COBRA obligations.)

Illegal or Wrongful Termination

Even a voluntary resignation can raise questions about whether or not the termination has some illegal basis. Here are some issues to consider:

- Did termination result from illegal discrimination based on non-job related issues such as race, sex, or age?
- Did the termination take place because the employee refused to do something illegal (such as falsify records or improperly bill a public agency)?
- Did the workplace environment cause the employee to resign?
- Is the employee being fired for exercising his or her protected rights, such as discussing unions or calling in an OSHA safety inspector?

Action Steps

1. Clean separation. When an employee leaves your organization, make sure all loose ends are resolved. This includes taking care of any payroll and benefits changes, retrieving company property from the employee, and trying to end the relationship on the best footing possible. Completing a termination checklist can help you make sure you don't forget anything. A checklist can help you avoid legal problems, too. A sample termination checklist appears on page 211 of Section III.

   The checklist can also structure your termination procedure. In general, you should plan to hold a termination meeting with the employee on or near the employee's last day on the job. Go through all the items on the checklist. A termination meeting does not have to be an adversarial confrontation; it's best to maintain a calm and professional demeanor. In some cases, the former employee may be a future customer or supplier, so you want to retain the employee's goodwill to your business. Words said in anger during a dismissal could come back to haunt you in a legal proceeding.
2. Avoid future liabilities. As noted above, there are several possible sources of future liability. Follow certain procedures and routines in order to avoid problems. Be sure to build these four routines into your termination procedures:
   1. Make sure you pay the departing employee correctly and include payment for leave, if applicable. State and local laws require departing employees to be paid within a certain period of time after termination.
   2. If applicable, offer the employee health insurance continuation in accordance with COBRA and COBRA notification procedures.
   3. Use an exit interview to detect any potential problems with a departing employee. If you have decided to terminate the employee, tell the employee the reason. Make sure to be consistent with the reasons in all documents or related forms detailing the cause of termination.
   4. Make sure the employee's termination does not violate any protections or laws. (See the next section.)
3. Make sure all your actions are documented, dated, and signed by the person creating the document.

Pre-Termination Review: Checklist

Before you accept or complete a termination, review the circumstances to make sure there are no inherent liabilities. Remember, legal problems can result from either a firing or a resignation that the employee feels forced into submitting. The latter would be considered a "constructive discharge" or, in effect, a firing.

The following checklist may help you review the termination to see whether there are any legal problems in termination. Answering "no" to any of the questions below indicates potential problems with the termination. You should consider delaying the termination until after the problem is clarified.

- Was the employee notified of the performance problems and given time to correct the problems?
- Is the termination based on job performance (as opposed to the employee's age, sex, race, religion, disability, etc.)?

- Is there sufficient documentation in place to justify the termination?
- Is the termination consistent with the way other employees have been treated under similar circumstances?

Answering "yes" to any of the following questions indicates potential problems with the termination. You should consider delaying the termination until after the problem is clarified.

- Is the employee being released because he or she called in regulators or inspectors to double-check your operations? That is, is the termination a reprisal against the employee?
- Is the employee being released for refusing to do something illegal?

Learn How to Improve Your Business

Try to learn why the separating employee wants to voluntarily leave your company. You may discover things that will help improve your company's operations and retain employees in the future.

An exit interview is one way to gather an employee's opinions of working for your company. A sample exit interview form appears on page 212 of Section III.

If the employee reports directly to you and you are the source of the problem, the person may not feel comfortable discussing his or her reasons for leaving. You may wish to have a neutral, third party conduct the exit interview instead.

USING A SIGNED SEPARATION AGREEMENT

Introduction

In some situations, such as the termination or layoff of a long-term employee, you may wish to use a separation agreement signed by the separating employee. In signing, the employee promises not to sue you for the termination. If you use such an agreement, you must offer the departing employee some consideration beyond whatever you are obligated to pay (that is, the employee's paycheck). Typical considerations include some kind of severance payment, the offer of outplacement assistance, and so on. For this consideration, the employee releases your company from further liability.

It is a good idea to consult with your lawyer before entering into any sort of separation agreement.

When to Consider an Agreement

This option really only needs to be considered when:

- The employee is an older, longer-termed employee (the law protects workers 40 years and older) and/or is in an executive-level position.
- The employee is being separated involuntarily (regardless of whether the reason is a layoff, downsizing, reorganization, or firing).
- If it is a firing, the reason is not outright gross negligence or some obvious form of criminal conduct.

Action Steps

If this option appears prudent, there are several things to do:

1. Consult an attorney to draw up the agreement.
2. Make sure the employee is notified of his or her rights under the Older Workers Benefit Protection Act and given ample time to consider the offer.
3. Implement a severance policy.

Severance Policy

A typical severance policy indicates the conditions under which a severance payment would be made (for employees in certain categories and under certain conditions) and how the severance amount is calculated (often, this might mean two weeks of pay for the first year of employment, and then one week of pay for each additional full and partial year).

Recordkeeping Requirements

INTRODUCTION

State and federal laws require that you keep various human resources and employment records. The requirements shown below describe federal requirements. Contact the labor or employment departments of the agencies in your state to determine any special or unique recordkeeping requirements in your state.

The personnel records form on page 205 of Section III can help you fulfill these requirements. Keep in mind that these recordkeeping requirements only apply if you are covered by the law in question.

Simplify

Establish a simple method for keeping records and assign that responsibility to someone. Effective recordkeeping can help in administration of your business and can help defend you against employment litigation, if it arises. It is also a good idea to keep any superceded policies as well as the current policies on file.

Policy/Act/Record	What to Keep	How Long to Keep It
Policy: At Will	Current policy statement and any superceded statement	Indefinitely
Policy: Equal Employment	Policy statement	Indefinitely
Equal Employment Opportunity Commission laws and regulations	EEO-1; keep a copy of most recent report filed	Indefinitely
	If a EEOC-related charge or action is brought against the employer, keep all personnel or employment records	Keep records until the final disposition of the charge or action is made
Affirmative Action Plan	Written affirmative action plan plus supporting records and superceded plans	No specified minimum; however, it is proposed that records be kept for two years from the date the record was made or the personnel action occurred, whichever is later, except for contractors with fewer than 150 employees or government contracts of less than $150,000.

Policy/Act/Record	What to Keep	How Long to Keep It
	Vietnam Era Veterans' Readjustment Assistance Act documents	One year after final payment under contract for: copies of reports given to state employment service about number of individuals hired during reporting period; number of disabled and non-disabled veterans; total number of disabled veterans hired; related documentation such as human resources records on job openings, recruitment, and placement. One year for all records of complaints and actions taken under the Act.
Family and Medical Leave Act	Basic payroll data, dates and hours of FMLA leave, copies of employee notices, premium payments, copy of employer leave and benefits policies and procedures, other records of any disputes, medical certifications, and history	Three years
Fair Labor Standards Act and Equal Pay Act	All records relating to wages, wage rates, job evaluations, payroll, individual contracts, job descriptions, merit and seniority systems, collective bargaining agreements, documents explaining pay differentials, and documents documenting exempt status	Three years
	Records relating to wage deductions, wage differentials, and work time schedules	Two years
	Certificates showing age	Keep until termination of employment
Age Discrimination in Employment Act	Documents containing employees' name, address, age, occupation, rates of pay, and compensation earned each week	Three years
Merit/Seniority Systems	Various documents	Keep records for full period that plan or system is in effect plus one year after plan or system is terminated.

(Continued)

Policy/Act/Record	What to Keep	How Long to Keep It
Independent contractors	No specific format required. However, internal tests, SS-8 forms, or copies of contracts would all apply. Copy of independent contractor's insurance information	All employment tax related documents should be kept for a minimum of four years from due date of tax
Exempt status	Test of exempt status	Three years
Tax withholdings	See "Payroll" in this section	Three years
COBRA	Notification forms	Three years
Employee benefits plans	Summary plan descriptions plus any related materials	Six years or as long as the plan remains in effect
Workers' Compensation		Consult state authorities
Unemployment insurance		See payroll records
Applicant flow	Hiring record or registration log; information on hiring procedures impacting women and minorities	Two years
Employment applications	Application forms	One year
Reference checks	Not specified	Not specified
Immigration Reform and Control Act of 1986	Employment eligibility verification form (also known as I-9 form)	Three years from date of hire or one year from date of termination, whichever is later
Safety	OSHA forms 301, 300, and 300A	Five years
Miscellaneous employee records	Personnel record form, performance reviews, etc.	See specific terms noted above. Otherwise, keep records for one year after termination

Sample Employee Handbook

Employee Handbook

INTRODUCTION

A typical employee handbook explains what you expect from your employees, how you will treat them, and what they will receive from you, the employer, in return. A well-written employee handbook can establish the basis for an effective and productive working relationship.

WHAT TO INCLUDE IN A HANDBOOK: GENERAL CONTENT

An employee handbook should contain information that describes your policies, guidelines, and/or rules for the working relationship. This includes what you expect from employees in terms of their working habits, job performance, personal conduct, and customer communications. You should also include information describing certain policies and practices for the workplace itself. For example, one important policy every employee should know about is the company's sexual harassment policy. Some states require that information about sexual harassment policies be communicated to employees. This communication can be efficiently included in an employee handbook. An employee handbook also should describe the benefits or other considerations (such as vacation leave) that employees can expect to receive from your company.

Finally, you may wish to include information about the history of your business, your current business and customers, and the products and services you provide.

WHAT TO INCLUDE: SPECIFIC ITEMS

Many employers wish to retain their flexibility to hire and fire employees easily and quickly. Such a position is achieved through an employment at-will personnel policy. That policy should be communicated to employees at the beginning of the relationship and is the basis of the following recommendations.

An employee handbook written from an employment at-will perspective should cover the following points:

- Include a prominent disclaimer near the beginning of the handbook that announces that employment is "at-will." The disclaimer should also assert your right as the employer to use your discretion and flexibility in administering and in modifying the policies and procedures contained in the handbook.
- Include a prominent disclaimer near the beginning of the handbook that states that the policies within the handbook are not intended to be and should not be construed as an express or implied contract.

- If you have a probationary (or orientation) period of employment, make it clear that during and after the orientation period, the employee does not enter into any kind of permanent employment arrangement and that the at-will employment relationship will not be modified.
- Avoid speaking or writing policies in absolutes or promissory terms. Do not use the words "never," "always," "will," "must," "permanent," "seniority," or "shall." Such terms imply that certain actions will invariably occur. This can deny you the flexibility and discretion you should retain in human resource management.
- Avoid the use of absolute rules, guidelines, and timelines unless absolutely necessary. Reserve the right to make exceptions.
- Do not limit your ability to terminate employment. Do not indicate that you will only fire an employee for "just cause." If you list grounds for termination, make it clear that the list is only partial and not all-inclusive, and that employees can be terminated for any reason.
- While you should indicate the benefits your company provides, be careful to note that the coverages indicated are not guaranteed indefinitely, and they may be changed at your discretion at any time. In general, avoid overly detailed descriptions of benefit coverages in the handbook. Where possible, use booklets or brochures supplied by your benefits provider to explain your benefits package.
- Include a receipt for employees to sign, date, and return to you acknowledging that they have received and have read a copy of the handbook. The receipt should state that the information contained in the handbook is provided for informational purposes and may be changed at any time. The signed receipt should be placed in the employee's personnel file.

THE SAMPLE EMPLOYEE HANDBOOK

The sample employee handbook covers all of the major areas discussed above. It also includes a disclaimer that gives you the right to change anything in your employee handbook at any time with or without notifying your employees and an accompanying form that documents employee receipt of the handbook.

Note: *While every business is unique, the policies included here are written for most employment situations. You should review and modify this information to suit the specific policies and procedures of your business before you distribute this handbook to your employees. Each state has its own laws and regulations regarding human resources issues. Some states may treat handbooks as contracts; others do not. Consult with your lawyer before adopting human resources policies and procedures. It is recommended that you ask your lawyer to review your employee handbook before distributing it to your employees.*

We've included a checklist of items on pages 117–119 of Section III to help you customize the sample employee handbook for your business.

Note: *Be sure to place the current date on the bottom of each page of your personnel handbook. As you change or modify a particular policy, change the date on the bottom of the page. This lets you know at a glance when your policy was last updated and when the modified policy went into effect.*

THE HANDBOOK ON CD

The CD enclosed with this handbook contains the text of the model employee handbook. You can save the text from the CD onto your hard drive, easily modify any handbook details, and then print out an employee handbook that is customized for your business. Again, because state and local laws may differ, it is always wise to have an attorney review any changes you make.

See the instructions on page xii for details.

Suggested Steps for Developing an Employee Handbook

STEP 1: DETERMINE WHAT YOU WANT

First determine what content you want to include in the handbook. Examine your business and determine what types of personnel policies and procedures would best suit your operations. You should only include policies and procedures you intend to follow.

STEP 2: OBTAIN PUBLICATIONS ON STATE EMPLOYMENT LAWS

All states have different employment laws regarding how employers may deal with their employees. It would be prudent to obtain a general publication of your state's employment statutes and regulations before you finalize your employee handbook. Some states require an employer to provide greater parental and family leave than the federal Family and Medical Leave Act provides. Some states have smokers' rights laws that impact an employer's adoption of workplace tobacco-use provisions. Some states protect on the basis of marital status or sexual orientation, which may impact the use of a nepotism policy. Other typical areas an employer may want to regulate at the employment site include the use of lie detectors, drug-free workplace policies, and time off for jury duty. Each of these differs by state.

STEP 3: DRAFT THE HANDBOOK

Once you've determined what human resources policies you want to include and how your state regulates the employment relationship, it's time to put those policies in print. The sample handbook in this section provides suggested policies to include in your handbook.

Here are some helpful tips for drafting the handbook:

- Keep the handbook as short as possible. You cannot regulate every situation that will occur.
- Use language that's easy to understand.
- Include only those policies and provisions you are willing to enforce.
- Eliminate all content that states that employees only will be terminated for good cause, as such statements appear to guarantee job security.

STEP 4: HAVE THE DRAFT REVIEWED BY AN ATTORNEY

After you have prepared the handbook draft, have it reviewed by an attorney who is familiar with your state's employment regulations.

STEP 5: TRAIN ALL SUPERVISORY AND MANAGEMENT STAFF ABOUT THE HANDBOOK

Train your supervisory and management staff on the book's policies and procedures, and how you expect them to be enforced. Keep the following points in mind during training:

- Instruct your managers and supervisors not to make remarks to employees about duration of employment or any other statements that alter the at-will employment relationship.
- Authorize only one person to make employment offers.
- Tell your managers and supervisors the name of the person in the company who is authorized to alter the handbook's policies and procedures. Be sure to include that information in the handbook, too.

STEP 6: DISTRIBUTE THE HANDBOOK TO YOUR STAFF

Make sure you have a handbook receipt signed by each employee.

STEP 7: REVISE THE HANDBOOK

The handbook is only as good as the policies and procedures it contains. You should periodically review the handbook and revise the policies as necessary to keep the publication up to date and in compliance with changing employment laws and changes in your operations.

[YOUR COMPANY]

Sample Employee Handbook

as of [fill in date] _____

Note: *Remember to place the current date on each page of your handbook. Change the date when you modify the policies.*

Table of Contents

Welcome to
[YOUR COMPANY]

[YOUR COMPANY] has prepared this handbook to provide you with an overview of [YOUR COMPANY]'s policies, benefits, and rules. It is intended to familiarize you with important information about [YOUR COMPANY], as well as provide guidelines for your employment experience with us in an effort to foster a safe and healthy work environment. Please understand that this booklet only highlights company policies, practices, and benefits for your personal understanding and cannot, therefore, be construed as a legal document. It is intended to provide general information about the policies, benefits, and regulations governing the employees of [YOUR COMPANY], and is not intended to be an expressly implied contract. These guidelines, however, are not intended to be a substitute for sound management, judgment, and discretion. [YOUR COMPANY] reserves the right to modify any of these policies during application as deemed appropriate.

It is obviously not possible to anticipate every situation that may arise in the workplace or to provide information that answers every possible question. In addition, circumstances will undoubtedly require that policies, practices, and benefits described in this handbook change from time to time. Accordingly, [YOUR COMPANY] reserves the right to modify, supplement, rescind, or revise any provision of this handbook from time to time as it deems necessary or appropriate in its sole discretion with or without notice to you.

No business is free from day-to-day problems, but we believe our personnel policies and practices will help resolve such problems. All of us must work together to make [YOUR COMPANY] a viable, healthy, and profitable organization. This is the only way we can provide a satisfactory working environment that promotes genuine concern and respect for others including all employees and our customers. If any statements in this handbook aren't clear to you, please contact [NAME], or [his or her] designated representative for an answer. This handbook supersedes any and all prior policies, procedures, and handbooks of [YOUR COMPANY].

NON-DISCRIMINATION POLICY

It is [YOUR COMPANY]'s policy to provide equal employment opportunity to all employees and qualified applicants without regard to race, color, religion, sex, age, national origin, marital status, Vietnam Era Veterans' status, or physical or mental disability, to the extent required by law. This policy applies to all personnel actions, benefits, terms, and conditions of employment including, but not limited to, hiring, placement, training, compensation, transfer, promotion, leave-of-absence, termination, layoff, and recall. It is [YOUR COMPANY]'s policy to prohibit any kind of harassment of employees, supervisors, or subordinates because of their race, color, religion, sex, age, national origin, marital status, Vietnam Era Veterans' status, or physical or mental disability, to the extent required by law. Violations of this non-discrimination policy should be brought to the attention of your supervisor or his/her supervisor if your supervisor is the subject of the complaint. Additionally, a violation of this policy may be made the subject of a complaint under [YOUR COMPANY]'s grievance procedure.

[YOUR COMPANY] considers the implementation and monitoring of this policy an important part of each supervisor's responsibility. Supervisors will inform all employees of our policy and shall take positive steps to seek adherence to the policy by all employees within the realm of their responsibility. Only [NAME] has the authority of [YOUR COMPANY] to modify and change the policies and guidelines as provided in this handbook.

The failure of any employee or supervisor to comply with this policy (to engage in any discriminatory or other inappropriate conduct) will be grounds for disciplinary action that may include termination of employment.

SEXUAL HARASSMENT POLICY

Sexual harassment is a violation of the law. [YOUR COMPANY] will not tolerate sexual harassment at any level by management or non-management employees. Therefore, any and all conduct of a sexual nature including words as well as physical acts that has the purpose or the effect of unreasonably interfering with an employee's work performance, or creates an intimidating, hostile, or offensive working environment is strictly prohibited. Requiring an employee to submit to sexually harassing conduct as a term or condition of employment is strictly prohibited.

Any employee confronted with a decision or behavior which he or she believes is contrary to the above policy should notify [YOUR COMPANY]'s president or his or her designated representative within 48 hours of the act. [YOUR COMPANY] will investigate the matter on a confidential basis and take appropriate action. Any employee who violates the above policy, or is found to have engaged in other inappropriate conduct, will be subject to appropriate disciplinary action, up to and including termination.

EMPLOYMENT ON AN AT-WILL BASIS

All employees of [YOUR COMPANY], regardless of their classification or position, are employed on an at-will basis, and their employment is terminable at the will of the employee or [YOUR COMPANY] at any time, with or without cause and with or without notice. No officer, agent, representative, or employee of [YOUR COMPANY] has any authority to enter into any agreement with any employee or applicant for employment on other than on an at-will basis and nothing contained in the policies, procedures, handbooks, manuals, job descriptions, application for employment, or any other document of [YOUR COMPANY] shall in any way create an express or implied contract of employment or an employment relationship on other than an at-will basis.

WORKING AND COMPENSATION
Attendance and Reporting to Work

Each employee is important to the overall success of our operation. When you are not here, someone else must do your job. Consequently, you are expected to report to work on time at the scheduled start of the work day or work shift. Reporting to work on time means that you are ready to start work, not just arriving at work, at your scheduled starting time.

[YOUR COMPANY] depends on its employees to be at work at the times and locations scheduled. Excessive absenteeism and/or tardiness is sufficient cause for termination of employment. The determination of excessive absenteeism will be made at the discretion of [YOUR COMPANY]. Absence from work for [indicate number: e.g., three] consecutive days without properly notifying your supervisor will be considered a voluntary resignation. After [indicate number: e.g., two] days absence, you will be required to provide a physician's certificate of disability to document an illness-related absence and to ensure that you may safely return to work.

[YOUR COMPANY] reserves the right to require a medical doctor's certification of illness or injury for any employee at other times upon request.

If you will be absent from the job for approved leave purposes (e.g., vacation or illness), you should notify your supervisor as far in advance as possible of your leave request. Otherwise, you are expected to notify your supervisor by the starting time of the work day that you will be late or absent and provide the reason for that absence. If your supervisor is not available, you should contact our main office by the starting time. Leave your number so that your supervisor can return your call. Failure to properly contact us may disqualify the employee from any sick leave payment for that day, and the day off will then be counted as an unexcused absence for disciplinary purposes. Your attendance record is a part of your overall performance rating. Your attendance may be included during your review and may be considered for other disciplinary action up to and including termination.

Where possible, medical and dental appointments should be scheduled around your assigned work hours; otherwise, they may be considered absences without pay. If you are unable to schedule an appointment before or after your shift, you are required to talk to your supervisor to make special arrangements.

Work Day Hours and Scheduling

The regularly scheduled work day for our business office is: Monday through Friday, [indicate starting time] to [indicate closing time]. The usual expected work day at jobsites is [indicate starting time] to [indicate closing time].

Particularly at jobsites, this regular schedule may vary depending on such factors as weather, materials supply, permit approval, etc. If you are unsure about expected starting times on any particular job assignment, ask your supervisor for clarification.

In case of unplanned conditions, such as bad weather, that may force a schedule change at the last minute, you should refer to [indicate policy or whom to contact].

For lunch or meals, our policy is:

- Employee meals will be [indicate time duration: 30 minutes, 1 hour, etc.].
- The meal period is [select one: paid, unpaid].
- All employees are required to take a lunch break and no employee is authorized, without prior supervisory approval, to perform work during the lunch period.

For rest periods or breaks, our policy is:

- Rest periods or breaks [select one: will, will not] be scheduled by [YOUR COMPANY].
- Scheduled breaks will be [indicate time duration: 10 minutes, 15 minutes, etc.].
- Scheduled breaks are [select one: paid, unpaid].

Note: *Refer to "Compensation Program and Administration" in Section I for suggested policies and possible regulations regarding meals breaks and rest periods. Also note that some states require breaks for defined consecutive periods of work.*

Note: *If you offer a flex-time working schedule, that policy should be explained. The following language may be used:*

Flex-Time Working Schedule

We offer a flex-time working schedule [**Note:** *if this is only available to certain workers, e.g. office staff, indicate that here*]. Under this schedule, employees still work the scheduled [indicate number] hours each day. However, employees have the option to report to work between the hours of [indicate time] and [indicate time] each day. Our core business hours, when every employee is expected to be on the job, are from [indicate time] to [indicate time] each day. Your full [indicate number] hour work day begins when you actually arrive on the job. For example, if you normally arrive at 7:30 a.m. but are 15 minutes late one day, your work day for that day begins at 7:45 a.m.

Recording Hours Worked to Payroll

You are expected to correctly record or note the times you report for and leave work each day. Use [indicate method: time sheets, time clocks, etc] for reporting your hours. Only you are authorized to [record or punch] your own time.

Note: *Explain any rules about the system used. You should require the employee to sign each record submitted to confirm that the time submitted is accurate.*

Pay Period and Payday

We pay employees [indicate frequency: weekly, biweekly, monthly, etc.]

Note: *Some states require that non-exempt employees be paid biweekly at a minimum. Check with your state agency.*

Payday is on [indicate a consistent payday: every Friday, or the 15th and 30th of each month, etc.].

A work week is defined as 12:01 a.m. Sunday through 12:00 a.m. Saturday.

Overtime

Occasionally it may be necessary for an employee to work beyond his or her normal work day hours. Overtime pay is paid only when work is scheduled, approved, and made known to you in advance by your supervisor. Under no circumstances shall an employee work overtime without the prior approval of his or her supervisor.

Hourly employees will receive overtime pay at a rate of one-and-one-half times their regular hourly rate for all hours worked in excess of 40 in a work week.

To the extent possible, overtime will be distributed equally among all employees in the same classification who are willing to work overtime, provided that the employees concerned are equally capable of performing the available work. Decisions in this area will be made by the [indicate position]. Any employee asked to work overtime will be expected to rearrange his/her personal schedule to work when requested, and, at the discretion of [YOUR COMPANY], will be required to work overtime when asked.

Note: *Employees who work overtime without prior approval are still required to be paid overtime wages under wage and hour laws.*

Holidays

We observe the holidays listed below. These holidays will be paid as long as the employee was present on the work days before and after that holiday, or had an acceptable excuse for being absent on either or both days.

Note: *Delete or add holidays according to your company's observance:*

New Year's Day	Good Friday	Labor Day	Thanksgiving
Martin Luther King Jr.'s Birthday	Memorial Day	Veteran's Day	Christmas
President's Day	Fourth of July	Columbus Day	

If a paid holiday falls within an employee's vacation period, it will not be considered a vacation day.

Note: *If you use an orientation period, include the following notice:*

Orientation Period

The first [indicate number of days: 90, e.g.] consecutive days of employment are considered an orientation period, which may be extended at the discretion of [YOUR COMPANY]. An employee may be terminated at any time during or after the orientation period at the sole discretion of [YOUR COMPANY]. At the end of the orientation period, your supervisor may evaluate and discuss your performance with you. The orientation period provides time for you to become familiar with your position and provides time for [YOUR COMPANY] to assess your ability to perform your job responsibilities and duties. This period does not modify the at-will status of your employment with [YOUR COMPANY].

Note: *Some employers don't grant vacation or sick leave during the orientation period.*

STANDARDS AND EXPECTATIONS FOR THE WORKPLACE

Safety

[YOUR COMPANY] believes in maintaining safe and healthy working conditions for our employees. However, to achieve our goal of providing a safe workplace, each employee must be safety conscious. We have established the following policies and procedures that allow us to provide safe and healthy working conditions. We expect each employee to follow these policies and procedures, to act safely, and to report unsafe conditions to his or her supervisor in a timely manner.

How to Report Unsafe Conditions or Practices

Employees are expected to continually be on the lookout for unsafe working conditions or practices. If you observe an unsafe condition, you should warn others, if possible, and report that condition to your supervisor immediately. If you have a question regarding the safety of your workplace and practices, ask your supervisor for clarification.

If you observe a coworker using an unsafe practice, you are expected to mention this to the employee and to your supervisor. Likewise, if a coworker brings to your attention an unsafe practice you may be using, please thank the employee and make any adjustments to what you are doing. Safety at work is a team effort.

Maintaining a Safe Worksite

We expect employees to establish and maintain a safe worksite. This includes but is not limited to the following applications:

- Maintaining proper fall-protection systems.
- Building and maintaining walkways, handrails, and guardrails.
- Properly lifting and lowering heavy objects.
- Inspecting tools and equipment for defects before use.
- Keeping walkways clear of debris.
- Construction and use of safe scaffolding.
- Inspecting, cleaning, and properly storing tools and equipment after use.
- Following established safety rules.

Using Safety Equipment

Where instructed or needed, [YOUR COMPANY] provides safety equipment and devices. You are required to use the equipment provided in the manner designated as proper and safe by the manufacturer. You will receive additional training on how to use safety equipment. Failure to properly use safety equipment may lead to disciplinary action, up to and including termination.

If you require safety equipment that hasn't been provided, contact your supervisor before performing the job duty.

Reporting an Injury

Employees are required to report any injury, accident, or safety hazard immediately to their supervisor(s). Minor cuts or abrasions must be treated on the spot. More serious injuries or accidents will be treated accordingly. Serious injuries must be reported on the injury or accident report form.

Note: *See the OSHA forms in Section III.*

Hazard Communications

Under OSHA's Hazard Communication Standard, you are entitled to know about any hazardous materials that you may come in contact with on the job. You are also entitled to receive information about those materials (provided on Material Safety Data Sheets or MSDSs), and to receive personal protective equipment for dealing with them. You are entitled to receive guidance and instruction about how to handle those materials safely. When you are working with employees from another company, you are also entitled to know about any hazardous materials those employees bring into your workplace.

[YOUR COMPANY] maintains a file and plan for hazard communications. This file contains the MSDSs and a list of the likely hazardous materials with which you may come into contact. That file is found on the worksite at [indicate location] and you may inspect it at any time.

You will receive training and education in these matters in the following way:

Note: *Indicate how and when employees will be trained; see "Hazard Communications" in Section I for suggestions.*

If you believe that you are dealing with a hazardous material and lack the appropriate information and/or safety equipment, contact your supervisor immediately.

Care of Equipment and Supplies

All employees are expected to take care of all equipment and supplies provided to them. You are responsible for maintaining this material in proper working condition and for reporting any unsafe or improper functioning of this material to your supervisor promptly.

Neglect, theft, and/or destruction of any assigned materials will be grounds for disciplinary action, up to and including termination.

Smoking at the Workplace

Note: *This item depends on two factors: state or local laws, and/or the business owner's preferences. Of the two factors, state or local laws take priority. Check with either your local governing body (city or county, mayor's office, city manager's office, etc.) or the appropriate state authority to see which laws apply in this area. If there are such laws, you may be able to obtain written instructions directly from the local governing body or state authority on what to do. These instructions may be used to form the language of your policy.*

If there are no governing laws in your area, then the issue becomes one of whether you wish to allow smoking on the job. This decision may apply somewhat differently to enclosed offices and construction sites. Nonetheless, given all the documented ways smoking negatively impacts employee health, productivity, equipment, facilities, and public reaction, a prudent course of action is simply to make your worksites smoke-free. The following policy assumes this general position:

It is the policy of [YOUR COMPANY] to provide smoke-free environments for our employees, customers, and the general public. Smoking of any kind is not permitted inside our office or on our worksites. Employees may smoke on scheduled breaks or during meal times, as long as the smoking is done outside the worksite or office. Excessive smoke breaks may mean that the employee will be expected to work longer during the day to make up for time lost smoking.

Violence and Weapons

[YOUR COMPANY] believes in maintaining a safe and healthy workplace, in part by promoting open, friendly, and supportive working relationships among all employees. Violence or threats of violence have no place in our business. Violence is not an effective solution to any problem. Employees are strictly prohibited from bringing any weapons, including knives, pistols, rifles, stun guns, Mace, etc., to the worksite or office. Neither threats of violence nor fighting will be tolerated.

Note: *Include the following section if you will use a grievance policy:*

If you have a problem that is creating stress or otherwise making you agitated, you are encouraged to discuss it with your supervisor. You are expected to immediately report any violation of this policy. Any employee found threatening another employee, fighting, and/or carrying weapons to the worksite will be subject to disciplinary action, up to and including termination.

Drug-Free Workplace

[YOUR COMPANY] does not tolerate the presence of illegal drugs or the illegal use of legal drugs in our workplace. The use, possession, distribution, or sale of controlled substances such as drugs or alcohol, or being under the influence of such controlled substances is strictly prohibited while on duty, while on [YOUR COMPANY] premises or worksites, or while operating [YOUR COMPANY]'s equipment or vehicles. The use of illegal drugs as well as the illegal use of legal drugs is a threat to us all because it promotes problems with safety, customer service, productivity, and our ability to survive and prosper as a business. If you need to take a prescription drug that affects your ability to perform your job duties, you are required to discuss possible accommodations with your supervisor. Violation of this policy will result in disciplinary action, up to and including termination.

This policy has the following implications:

1. All employees are prohibited from unlawfully manufacturing, distributing, dispensing, possessing, or using controlled substances in the workplace. Controlled substances include (but are not limited to):

   ■ Narcotics (e.g., heroin or morphine)
   ■ Cannabis (e.g., marijuana)
   ■ Stimulants (e.g., cocaine)
   ■ Depressants (e.g., tranquilizers)
   ■ Hallucinogens (e.g., PCP or LSD)

2. Any employee who violates this policy will be subject to disciplinary action, up to and including termination.
3. Employees are encouraged to seek help and assistance for any drug abuse problem by taking advantage of any educational or rehabilitation services we provide.
4. Your receipt of this policy statement and signature on the handbook acknowledgment form signify your agreement to comply with this policy.
5. Any employee convicted of violating criminal drug statutes in this workplace must notify an appropriate officer or senior official of [YOUR COMPANY] of that conviction within five days of the conviction. Failure to do so may lead to disciplinary action. Employees convicted of violating criminal drug laws may be given the option of participating in an approved rehabilitation program.

Moonlighting

[YOUR COMPANY] discourages our employees from taking additional outside employment. If an employee is planning to take an outside job, permission must first be obtained from his or her supervisor and the president. Work requirements for [YOUR COMPANY], including overtime, will take precedence over any outside employment.

Permission will not be given for an employee to take any outside job in the employ of a company which is in the same or related business as [**YOUR COMPANY**], or which is in any way a competitor of [**YOUR COMPANY**].

If permission is granted for a person to take outside employment, the employee must report to his or her supervisor when the outside job has started. If the employee is unable to work when requested by [**YOUR COMPANY**], including overtime, or is unable to maintain a high work performance level at [**YOUR COMPANY**] as a result of moonlighting, permission to work at the outside job may be rescinded, or the employee may be subject to dismissal.

[**YOUR COMPANY**] will not pay medical benefits for injuries or sickness resulting from employment by any other employer.

LEAVES AND ABSENCES

Note: *For guidelines, see the section on "Pay for Time Not Worked" under "Benefits Administration" in Section I.*

Sick Leave

A full-time employee is granted [indicate number] hours of sick leave hours after [indicate number] months of continuous employment with [**YOUR COMPANY**]. Thereafter, the employee will be granted [indicate number] hours of sick leave credit for each additional month of regular employment. The total amount of unused sick leave may be carried over from one year to the next [indicate any applicable limits, such as the maximum number of days or hours allotted for sick leave].

See the section on "Attendance and Reporting to Work" for details about how to report in on days when you are sick.

A part-time employee [**Note:** *Select one of the following options*]

receives the same benefits as a full-time employee but at [indicate prorated amount: 50 percent, e.g.] of full-time benefits based on the number of hours worked.

[OR]

is not eligible for sick leave hours.

Sick leave is a benefit granted to employees as a protection during a period of unexpected illness or injury. It is not earned as an entitlement. Therefore, the employee will not be paid for any unused, accumulated sick leave hours at the time of termination.

Vacation Leave

[**YOUR COMPANY**] provides its employees a vacation benefit each year as a way to express our appreciation and a way to renew and refresh our employees. Because our business is often very seasonal, [**YOUR COMPANY**] reserves the right to grant vacation leave at times that are most suitable for our business conditions and to limit vacation during our busy season.

A full-time employee becomes eligible for vacation leave after [indicate number] months of continuous employment with [**YOUR COMPANY**]. The following schedule for vacation is listed below. **Note:** *These are examples only, not vacation benefit suggestions.*

Length of Employment	Amount of Vacation Benefit
Up to 3 years	10 days
At 3rd anniversary	15 days
At 10th anniversary	18 days
At 15th anniversary	20 days

Vacation leave should generally be taken in the year it is granted. It should be scheduled and approved by [**YOUR COMPANY**] at least two weeks in advance.

Any accumulated but unused vacation leave will be paid to the employee upon termination if proper notice of termination is given. Unused vacation leave will not be paid if an employee is terminated for misconduct or gross misconduct of any kind.

A part-time employee [**Note:** *Select one of the following options:*]

receives the same benefits as a full-time employee but at [indicate prorated amount: 50 percent, e.g.] of full-time benefits based on the number of hours worked.

[OR]

is not eligible for vacation leave hours.

Personal Leave

After [indicate number] months of continuous employment with [**YOUR COMPANY**], an employee is eligible to receive [indicate number] days of personal leave each year. Personal leave can be used for any purpose, including personal business or emergencies. Personal leave days may not be used on the days before or after paid vacation days or holidays. Whenever possible, the employee should schedule and receive approval for personal leave at least three days in advance. Unused personal leave may not be accumulated from one year to the next. An employee will not be paid for any unused, accumulated personal leave at the time of termination.

A part-time employee is not eligible for personal leave.

Other Leave

Bereavement Leave

[**YOUR COMPANY**] will provide time off for the employee upon the death of an immediate family member. Employees who have been with the [**YOUR COMPANY**] for [indicate number] months continuous of employment are eligible to receive up to three days of paid leave a calendar year. Other accommodations may be made depending on the circumstances. Notify and confer with your supervisor or other senior official of [**YOUR COMPANY**]. Appropriate documentation of this event may be necessary to support unusual requests.

Members of the employee's immediate family include spouse, parents, children, siblings, grandparents, and in-laws.

Leave will be paid at straight time for the hours the employee was scheduled to work on the days missed.

Jury Leave

Employees who are called for jury duty will be granted time off with pay to perform this civic duty. You must provide your supervisor with the jury summons notice and a note from the Clerk of the Court indicating the times you were at the court for jury duty. [YOUR COMPANY] will pay you straight time for your regularly scheduled hours of work, minus the compensation you received from the court for your service for up to [indicate time: 5 days, 2 weeks, etc.]. If you are excused from jury duty on any day or any portion of a day, you are expected to report for work for the remainder of that day, or otherwise notify your supervisor.

Military Leave

Employees called into military service will be granted an unpaid leave of absence and reemployment rights as provided by the laws of the United States. Appropriate military papers must be presented to [YOUR COMPANY] before you are granted military leave of absence. You may use vacation and/or personal leave to compensate during this period. However, this is not required.

Family and Medical Leave

Any employee who has been employed by [YOUR COMPANY] for 12 months and who worked at least 1,250 hours during the previous 12 months will be permitted to take 12 weeks of unpaid leave during a 12-month period in the following situations:

- The birth or adoption of a child in the family.
- The employee or a member of his/her immediate family (spouse, child, or parent) experiences a serious illness. A serious illness is an illness, injury, impairment, or physical or mental condition that involves either in-patient care in a hospital, hospice, or residential medical care facility.
- A serious illness requiring continuing treatment by or under the continuing supervision of a health care provider (a state licensed doctor or osteopath, or a federally approved person capable of providing health care services).
- Pregnancy.

Any employee needing leave as a result of the birth or placement of a child is required to provide at least 30 days notice before the commencement of the leave or as much notice as is feasible. Leave for the birth or placement of a child can be taken only within 12 months of the birth or placement. If you and your spouse both want to take leave for the care of a newly arrived child or a sick parent, the aggregate leave is limited to 12 weeks. If the leave is requested to care for a sick child or a spouse, each spouse is entitled to 12 weeks of leave.

Any employee needing leave because of the serious health condition of a spouse, child, or parent or his or her own serious health condition is required to provide at least 30 days notice before the commencement of the leave or as much notice as is feasible. If leave is requested for a personal serious health condition or that of a family member, you will be required to provide certification from the health care provider supporting the need for the requested leave in a timely manner. The certification must contain the following information:

- The date on which the serious health condition began.
- The probable duration of the condition.
- The medical facts regarding the condition.
- A statement that you are unable to perform your functions or that you are needed to care for a spouse, parent, or child.
- An estimate of the time required to recover or care for a spouse, parent, or child.
- In the case of intermittent leave, the date and duration of treatment.

[YOUR COMPANY] may require a second opinion, at its own expense, from an independent health care provider designated or approved by [YOUR COMPANY]. If the second opinion differs from the original certification you provided, [YOUR COMPANY] may obtain a third opinion, at its expense, from a mutually acceptable health care provider, whose opinion is final and binding on you and [YOUR COMPANY]. Recertification may, at the option of [YOUR COMPANY], be required on a reasonable basis.

Employees desiring leave for a personal serious health condition or a serious health condition of a spouse, child, or parent may take leave on an intermittent or reduced schedule if medically necessary. If such leave is for planned medical treatment, you must make a reasonable effort to schedule the treatment so as not to unduly disrupt [YOUR COMPANY]'s operations and give 30 days advance notice or as much notice as is feasible. It may be necessary to require you to temporarily transfer to an alternative position with equivalent pay and benefits if the transfer better accommodates your recurring periods of leave.

Employees desiring family and medical leave under this section must first use their paid vacation, sick leave, and other paid leave. If available paid leave is less than 12 weeks, unpaid leave will be made available to complete the 12 weeks of family and medical leave. Your group health benefits will continue during your leave; however, you must continue to pay your portion of the premiums. If you fail to return from leave for reasons other than the continuance, recurrence, or onset of a personal serious health condition or a serious health condition of a family member, you will be required to reimburse [YOUR COMPANY] for any premiums it paid on your behalf to maintain your health coverage during your unpaid leave period. You will not lose your previously accrued employment benefits as a result of taking a leave; however, benefits, other than group health insurance, will not accrue during the leave period.

Upon completion of your Family and Medical Leave, you will be returned to your former position or to a position equivalent in pay, benefits, and other terms and conditions of employment. If you are one of the highest paid ten percent of [YOUR COMPANY]'s workforce within a 75-mile radius of your worksite, you may be denied restoration of employment if your return would cause substantial and grievous economic injury to [YOUR COMPANY]'s business. In such a case, you will be notified of [YOUR COMPANY]'s intent not to restore your employment at the time the determination is made and will be given an opportunity to return to work at that time. If you elect not to return after such notice, [YOUR COMPANY] will continue to pay its portion of the premiums for your health benefits until the end of your leave.

CUSTOMERS AND COMMUNICATIONS

Answering the Telephone

Our current and potential customers often call us. The impression we make when answering the phone can be very important to the success of our business. Therefore, you should always answer the phone with a pleasant greeting, stating [YOUR COMPANY]'s name and introducing yourself:

"Hello, [YOUR COMPANY], this is (Employee's name). How can I help you?"

Listen carefully to the caller's request. Make every effort to be helpful in meeting the caller's needs and requests.

Responding to Customer Inquiries and Problems

Note: *If you have a policy on customer satisfaction, include it here.*

Confidentiality of Company, Customer, and/or Supplier Information

During the course of your duties, you may come across information about [YOUR COMPANY], our customers, suppliers, and/or about other employees. In general, unless this information is publicly known, employees should consider that it is sensitive, private, and confidential. Do not reveal this information to any other employee, co-worker, supplier, or the public without the express consent or under the explicit direction of your supervisor or a senior official of [YOUR COMPANY].

Furthermore, make sure that any materials containing such information are filed and/or locked up before leaving your work area every day. During the work day, do not leave any sensitive information lying about or unguarded.

If you have any questions about this policy, consult your supervisor.

PERSONAL CONDUCT

Appearance and Dress

To present a business-like, professional image to our customers and the public, all employees are required to wear appropriate clothing on the job. By necessity, the dress standards for the business office are somewhat different than for jobsites.

- For the business office, casual to business-style dress is appropriate. Employees should be neatly groomed and clothes should be clean and in good repair. Leisure clothes such as jeans, cut-offs, or halter tops are not acceptable attire for the business office.
- For jobsites, employees are expected to wear work clothes appropriate for work to be done. Employees should be sensitive to the location and context of their work and should be ready to adjust their dress (for example, by wearing shirts) if the circumstances so warrant. Employees at a jobsite should wear clothing that protects their safety (steel-toed shoes, for example) and wear clothing in such a way as to be safe (e.g., shirts tucked in when working around machinery).

Conflict of Interests

You should avoid external business, financial, or employment interests that conflict with [YOUR COMPANY] or with your ability to perform your job duties as described. A conflict of interest exists when you stand to gain through another party who acts in a way to harm your current employer. This applies to your possible relationships with any other employer, consultant, contractor, customer, or supplier.

Violations of this rule may lead to disciplinary action, up to and including termination.

Note: *Some states may have laws limiting how much of a restriction you may place on employees about working elsewhere. Check with the appropriate state authority for guidance.*

Code of Ethical Conduct

In order to avoid any appearance of a conflict of interest, employees are expected to abide by the following code of ethical conduct. Please consult your supervisor or an official of [YOUR COMPANY] if you have any questions.

Employees of [YOUR COMPANY] should not solicit anything of value from any person or organization with whom [YOUR COMPANY] has a current or potential business relationship.

Employees of [YOUR COMPANY] should not accept any item of value from any current or potential business party in exchange for or in connection with a business transaction between [YOUR COMPANY] and that other party.

Employees may accept items of incidental value (generally, no more than $25) from customers, suppliers, or others as long as the gift is not given in response to solicitation on your part and as long as it implies no exchange for business purposes. Items may include gifts, gratuities, food and drink, and/or entertainment.

If you are faced with a situation that violates this code of ethic conduct, notify your supervisor.

Solicitation and Distribution

For the safety, convenience, and protection of all employees, [YOUR COMPANY] has adopted the following rules concerning solicitation and the distribution of materials:

- Persons who are not employed by [YOUR COMPANY] are not permitted to solicit or distribute materials for any purpose on [YOUR COMPANY] property or jobsites.
- Solicitation by employees for any purpose is strictly prohibited during working time, including the working time of the employee who is soliciting and the employee being solicited. This prohibited time does not include time spent by employees for rest periods, meal periods, or other free periods during the day when employees are not supposed to be performing job tasks.
- The circulation or distribution of literature or written materials of any description for any purpose is strictly prohibited in working areas at all times.

Personal Calls, Visits, and Business

We expect your full attention while you are working. We realize that you may have to take care of some personal matters, such as checking up on your children or making arrangements. If you must do such things, please try to make calls or conduct personal business during breaks or meal periods. Regardless of when the call is made, it should be kept short.

In the same context, please limit incoming personal calls, visits, or personal transactions. Our phones should be available to serve our customers and non-business use of the phones can hurt our business. A pattern of excessive personal phone calls, personal visits, and/or private business dealings is not acceptable and may lead to disciplinary action.

Business Expenses

You may incur expenses while working for [YOUR COMPANY]. We will reimburse you for typical business expenses, such as:

- Mileage, when we ask you to go to a different worksite for a special reason.
- When you must purchase supplies or materials out of your own pocket at the request of your supervisor. Your supervisor must provide you with a written note approving the expense prior to the purchase.

You should retain a receipt if possible. Claims for reimbursement should be filed within 45 days of when they were incurred. Reimbursements will be processed and paid according to [YOUR COMPANY] policy and procedures.

Privacy and Company Property

As an employee, you use the property and equipment [YOUR COMPANY] owns and provides. You may also use [YOUR COMPANY]'s materials, information, and other supplies. While you may decorate your office workspace with your own personal possessions (such as pictures, plants, and the like), remember that supplied property belongs to [YOUR COMPANY]. We reserve the right to search any such property (e.g., personal computers, desks, lockers, or other storage areas) at any time. We also reserve the right to inspect personal property (e.g., tool boxes, purses, briefcases) as employees leave their worksites. Refusal to allow inspection may lead to disciplinary action, up to and including termination.

If a search is deemed necessary, it will generally be done on a non-discriminatory basis for legitimate business reasons. This includes situations in which there is reasonable concern that there may be a violation of [YOUR COMPANY]'s rules and policies.

Electronic Communications

Computers, software, fax machines, and related equipment are for work purposes, not personal usage. [YOUR COMPANY] reserves the right to monitor and review all incoming and outgoing electronic communications with or without notifying employees. In addition:

- You are responsible for keeping your individual password secure. You may not permit any other person to use your account and/or password.
- You may not use another person's password to download, transmit, or read information.
- Electronic communications downloaded, transmitted, or in your possession may not contain content that could be reasonably considered offensive, pornographic, or disruptive to other employees. Offensive content would include, but is not limited to, sexual comments or images, racial slurs, gender-specific comments, or any comments that would offend someone on the basis of his or her age, sexual orientation, religious or political beliefs, national origin, race, gender, or disability.
- The use of electronic communications to conduct non-work related private businesses (enterprises), participate in pools, gamble, or support personal political causes or activities is prohibited.

Confidentiality

All company records are confidential. No records of any kind nor any information regarding company matters or customers may be released to anyone outside the company. Violation of the confidentiality policy is grounds for immediate discharge.

Maintaining Your Personnel Records

It is your responsibility to provide current information regarding your address, telephone number, insurance beneficiaries, change in dependents, marital status, etc. Please use the personnel records form to note any changes in your address, phone number, emergency contact information, marital status, number of dependents, etc. Changes in exemptions for tax purposes will only be made upon the receipt of a completed W-4 form.

Viewing Your Personnel Records

You may examine the contents of your personnel folder by appointment during regular business hours. You can review the contents of your folder no more than once a quarter. Since the contents of the folder are [YOUR COMPANY]'s property, you cannot keep these materials. A [YOUR COMPANY] official will be present while you review your personnel folder.

If you notice something incorrect in your folder, you should complete a Personnel Records Form and submit it to the appropriate official. That person will review the form and make any appropriate changes. If you wish to contest information in the folder, submit a written note that describes the item(s) you wish to contest, why it should be changed, and the way you believe it should be written. After the note is reviewed, changes will be made to the original document or your note will be attached to the unchanged document.

Note: *Some state laws may have regulations concerning employee access to personnel records and other related rights. Consult with the appropriate state agency to determine any specific requirements of your state.*

BENEFITS

In addition to the leave policies previously noted, [YOUR COMPANY] offers the following benefits to employees:

Note: *Briefly list any benefits you offer.*

You will be provided with descriptive information about those benefits separately.

COBRA Notification

If you become covered by [**YOUR COMPANY**]'s group health insurance program, you have a right under federal law to continue your coverage under the plan upon:

■ Termination of your employment for reasons other than gross misconduct.
■ A reduction in your hours of employment that results in the loss of coverage under the plan.

The spouse of an employee covered by [**YOUR COMPANY**]'s group health insurance program has the right to choose continuation coverage for himself or herself if he or she loses coverage under the plan for the following reasons:

■ Death of the employee.
■ Termination of the employee's employment for reasons other than gross misconduct or a deduction in the employee's hours resulting in the loss of coverage.
■ Divorce or legal separation from the employee.
■ The employee becomes entitled to Medicare.

A covered retiree or a covered retiree's widowed spouse has a right to choose continuation coverage for themselves and their dependents in the event of a bankruptcy proceeding under Title II that results in the substantial elimination of their coverage within one year before or after the filing of the bankruptcy, provided the retiree retired on or before the date coverage is eliminated.

A dependent child of an employee covered by [**YOUR COMPANY**]'s group health insurance program has the right to continuation coverage if group coverage under the plan is lost because of:

■ The death of the employee.
■ The termination of the employee's employment for reasons other than gross misconduct or a reduction in the employee's hours of employment.
■ The employee's divorce or legal separation.
■ The employee becomes entitled to Medicare.
■ The dependent ceases to be a dependent child under the [**YOUR COMPANY**]'s program.

Under Federal law, the employee or a family member must notify [**YOUR COMPANY**] within 60 days of the occurrence of a divorce, legal separation, or a child losing dependent status under [**YOUR COMPANY**]'s program. [**YOUR COMPANY**], in turn, must notify the insurance carrier of the employee's death, termination of employment, reduction in hours, or Medicare entitlement.

When [**YOUR COMPANY**] receives notification that one of these events has happened, it will notify you that you have the right to continuation coverage. You, in turn, must notify [**YOUR COMPANY**] within 60 days from the date you lose coverage that you want continuation coverage. If you do not choose continuation coverage, your group health insurance coverage will end.

If you choose continuation coverage, [**YOUR COMPANY**] will give you coverage that is identical to the coverage provided under its health insurance program to similarly situated employees or family members. You will be given the opportunity to maintain coverage for three years unless you lose your coverage because of:

■ A termination of employment or reduction in hours, in which case the period is 18 months.
■ If [**YOUR COMPANY**] files for bankruptcy and you qualify as a retiree or retiree's widow, in which the continuation period is life coverage.

Your continuation coverage may be cut short for any of the following reasons:

■ [**YOUR COMPANY**] no longer provides group health insurance coverage to any of its employees.
■ You don't pay the premium for your continuation coverage.

- You become covered under another group's health plan.
- You become entitled to Medicare, except where [**YOUR COMPANY**]'s filing for bankruptcy is the qualifying event.

You do not have to show you are insurable to choose continuation coverage. However, you are required to pay all of the premiums for the continuation coverage plus a fee equal to two percent of the premium. When your continuation coverage period ends, you will be allowed to enroll in an individual conversion health plan offered by the insurance company providing [**YOUR COMPANY**]'s group plan. You may have additional rights under [indicate your state] law. If you have any questions concerning your rights, please contact [assign a position]. Also notify [assign a position] of there is a change in your marital status or if you or your spouse's address has changed.

EMPLOYEE MANAGEMENT POLICIES

Grievance Problem-Solving

If you experience any problems or concerns relating to your job performance or work situation, you are encouraged to discuss them with an appropriate official of [**YOUR COMPANY**] so that the problem can be addressed and resolved most effectively. Any employee who brings forward such problems or concerns in a good-faith manner will not be penalized for raising these matters. While it may not be possible to guarantee your confidentiality while addressing the problem, every effort will be made to protect employee privacy.

[**YOUR COMPANY**] will make every effort to address the problem quickly and thoroughly. Where possible and appropriate, we will take appropriate action to resolve the problem fairly. You will be informed of any outcome and the reason for it.

The following procedures should be followed:

1. If you are experiencing some kind of problem relating to your work, discuss the problem with your immediate supervisor if possible. If that is not possible, you may ask to discuss the problem with some senior official of [**YOUR COMPANY**] if you are unable to discuss it with your supervisor.
2. If you do not receive a response to your problem from your immediate supervisor within a reasonable time, you should bring this matter to the attention of the president of [**YOUR COMPANY**].
3. If you are not satisfied with the response to your problem, you may ask that response be reviewed and reconsidered by a senior official of the business, including the president. The president's decision will be final.

Performance Evaluations

Employees may have their job performance reviewed:

[indicate when]
[indicate what procedure/form will be used]

Progressive Disciplinary Procedure

You may receive a written warning if you violate [**YOUR COMPANY**]'s employment policies or general work rules and regulations unless the seriousness of the violation calls for a stronger disciplinary action. Documentation of all disciplinary actions, including verbal and written warnings, will become a permanent part of the employee's personnel file. Receipt of two written warnings in any 12-month period is cause for suspension without pay of up to five days. Receipt of three written warnings in any 12-month period will result in further disciplinary action, up to and including dismissal.

[**YOUR COMPANY**] will not re-employ anyone who was dismissed from its employ.

Disciplinary Action

It is not [YOUR COMPANY]'s sole intent to punish, but rather to correct and impress upon the employee the serious consequences of his or her unacceptable performance or undesirable or unacceptable conduct. To accomplish that goal, [YOUR COMPANY] uses a progressive disciplinary process that includes verbal warnings, written warnings, and separation. Based upon a comprehensive review of the circumstances and the nature of the undesirable or unacceptable conduct, [YOUR COMPANY] will decide which form of discipline will be used. It is within [YOUR COMPANY]'s sole discretion to select the appropriate disciplinary action to be taken among the various disciplinary options outlined below. Notwithstanding the availability of the various disciplinary options, [YOUR COMPANY] reserves the right to separate an employee from its employ at its discretion, with or without notice.

[YOUR COMPANY] has available the following disciplinary options for the administration of corrective action for violation of its rules and undesirable or unacceptable work behavior other than gross misconduct:

- Verbal warning
- Written warning
- Suspension without pay
- Dismissal

Gross Misconduct

Gross misconduct includes, but is not limited to: insubordination; theft; dishonesty; excessive absenteeism; unauthorized use of the name or property of [YOUR COMPANY]; breach of confidentiality; possession of firearms, other weapons, and/or illicit substances or alcohol on [YOUR COMPANY]'s property or worksites; malicious and willful damage to property belonging to [YOUR COMPANY] or its customers; reporting to work under the influence of alcohol or any illicit substance; failure to be attentive to safety issues relating to [YOUR COMPANY]; failure to report an employee accident within the same day of the occurrence; falsification of an employment record or time card; punching another employee's time card; fighting on [YOUR COMPANY]'s property or worksite; unauthorized departure from a worksite; gross negligence; or any other conduct that is deemed detrimental to the public image and reputation of [YOUR COMPANY].

This list is not all-inclusive of behavior that may be deemed gross misconduct. No disciplinary action other than dismissal will be considered for any conduct considered by [YOUR COMPANY] to be gross misconduct.

Re-Employment

Former employees will not be rehired if they:

- Were dismissed by [YOUR COMPANY]
- Resigned without giving proper notice
- Were dismissed for inability to perform job duties
- Had a poor attendance record
- Had a below-average evaluation
- Violated work rules or safety rules

Former employees who are rehired and return to work within three months of their termination won't be required to go through another orientation period, unless [YOUR COMPANY] deems it necessary. Former employees who are rehired and return to work more than three months after their termination will be rehired only as new employees and must complete a new orientation period. They will be considered new employees for any and all benefits.

Acknowledgement of Receipt of Employee Handbook

I have received the current [YOUR COMPANY] employee handbook and have read and understand the material covered. I have been allowed to ask questions, and realize that [assign a staff member] and/or a designated representative will clarify the covered material, should I require it. I understand that any future questions that I may have about the handbook or its contents will be answered by [assign a staff member] or his or her designated representative upon request. I agree to and will comply with the policies, procedures, and other guidelines set forth in the handbook. I understand that [YOUR COMPANY] reserves the right to change, modify, or abolish any or all of the policies, benefits, rules, and regulations contained or described in the handbook as it deems appropriate at any time, with or without notice. I acknowledge that neither the handbook nor its contents are an expressed or implied contract regarding my employment.

I further understand that all employees of [YOUR COMPANY], regardless of their classification or position, are employed on an at-will basis, and their employment is terminable at the will of the employee or [YOUR COMPANY] at any time, with or without cause, and with or without notice. I have also been informed and understand that no officer, agent, representative, or employee of [YOUR COMPANY] has any authority to enter into any agreement with any applicant for employment or employee for an employment arrangement or relationship other than on an at-will basis and nothing contained in the policies, procedures, handbooks, or any other documents of [YOUR COMPANY] shall in any way create an express or implied contract of employment or an employment relationship other than one on an at-will basis.

This handbook is [YOUR COMPANY]'s property and must be returned prior to separation.

_____   _____
Signature   Date

_____
Employee Name: Printed

Personnel
Management
Documents

Employee Handbook Customization Checklist

You should customize the sample employee handbook to suit your business. The topics shown below indicate the various sections of the handbook that may require modification by either picking one option of several, or by providing specific details. Check off each section as you complete the customization.

Issue or Policy Needing Modification	Section	Change Made	Used/ Not Used	Date Put Into Effect
Working and Compensation				
Company Attendance Policy	Attendance and reporting to work			
Regularly scheduled work day hours	Work day hours and scheduling			
Lunch or meal breaks	Work day hours and scheduling			
Rest periods	Work day hours and scheduling			
Explaining any flex-time schedule in use	Work day hours and scheduling			
Reporting hours worked for payroll purposes	Recording hours worked to payroll			
When employees are paid	Pay period and payday			
Special pay rates for unusual work periods	Overtime			
Holidays observed; pay policies for holidays	Holidays			
New employee orientation period	Orientation period			
Standards and Expectations				
Company safety policy	Safety			

Issue or Policy Needing Modification	Section	Change Made	Used/ Not Used	Date Put Into Effect
Standards and Expectations *(Continued)*				
Location of material safety data sheets	Hazard communications			
Equipment policy	Care of equipment and supplies			
Smoking policy	Smoking in the workplace			
Resolving stressful situations	Violence and weapons			
Drug Policy	Drug-free workplace			
Policy for employees holding outside jobs	Moonlighting			
Leave and Absences				
Type of sick leave to offer	Sick leave			
Part-time employee sick leave	Sick leave			
Type of vacation leave to offer	Vacation leave			
Part-time employee vacation leave	Vacation leave			
Type of personal leave to offer	Personal leave			
Types of provisions to include for other forms of leave	Other leave			
Customers and Communication				
Company phone policy	Answering the telephone			
Customer satisfaction policy	Responding to customer inquiries and problems			
Maintaining company privacy	Confidentiality of company, customer, and/or supplier information			

Issue or Policy Needing Modification	Section	Change Made	Used/ Not Used	Date Put Into Effect
Personal Conduct				
Dress policy	Appearance and dress			
Employee conflict of interest	Conflict of interests			
Employee acceptance of gifts	Code of ethical conduct			
Non-employee/ employee non-work related solicitation	Solicitation and distribution			
Personal business during work time	Personal calls, visits, and business			
Claim deadline for business expense reimbursement	Business expenses			
Company access to employee areas/ possessions	Privacy			
Employees reviewing their personnel files	Viewing personnel records			
Benefits				
List of benefits provided	Benefits			
Employee Management Policies				
Employee grievance policy	Grievance problem-solving			
Policy and procedure for performance appraisals	Performance evaluation			
Company disciplinary actions	Progressive disciplinary procedures			
Company rules	General work rules and regulations			
Rehiring former employees	Re-employment			

Decision Planning Worksheet
and Checklist

Use this tool to keep track of your decisions relating to human resources policies and procedures used to develop your own handbook and the status of their implementation.

Issue	Recommended Action	Status and Date(s) Policy(ies) Drafted/Adopted
Employment At-Will	Adopt employment at-will as your company's policy and practice. Communicate policy in: ■ Employee handbook ■ New employee orientation	
Equal Employment Opportunity	Adopt policy. Communicate policy in: ■ Employee handbook ■ Posters	
Affirmative Action	Develop, write, and implement plan.	
Family and Medical Leave	Provide proper notices. Two-day response procedure.	
Pregnancy	Policy on temporary disability should cover pregnancy.	
Employing Persons with Disabilities	Identify essential functions of job in job descriptions. Use job-related selection procedures; consider making reasonable accommodations. Have any medical exam done *after* conditional job offer. Keep medical records confidential and separate from personnel records.	

Issue	Recommended Action	Status and Date(s) Policy(ies) Drafted/Adopted
Drug-free Workplace Job	Publish policy on drug use: ▪ To every employee ▪ Employee handbook ▪ Employee training	
Descriptions	Develop descriptions for each job, with essential functions noted. Provide each employee with a job description.	
Independent Contractors	Make sure anyone working as an independent contractor meets the criteria. Use a written contractual agreement signed by all parties.	
Exempt Employees	Make sure "exempt" employees meet the criteria. Keep "test" records on file.	
Compensation Administration	Make sure you are paying employees properly: ▪ Minimum wage ▪ Equal pay for equal work ▪ Overtime ▪ Rest breaks Your work week starts on: _____	
Payroll Records	Keep accurate records of wages and hours worked for each employee.	
Workers' Compensation	Make sure you are in compliance.	
Recruiting	Keep ads non-discriminatory.	
Applications for Employment	Use a legally acceptable and effective application form.	

Issue	Recommended Action	Status and Date(s) Policy(ies) Drafted/Adopted
Applications for Employment *(Continued)*	Have a clear policy on when you will accept applications and how long you will maintain applications on file. Keep records of job applicants and candidates who were hired.	
Candidate Selection	Use strong, job-related selection techniques.	
Checking References	Check references thoroughly and appropriately.	
Offer of Employment Letter	Use a written letter offering employment.	
Employment Eligibility	Follow the I-9 procedure for verifying new employee eligibility to work.	
New Employee Orientation	Follow a set procedure for inducting all new employees into your business.	
Safety Policies and Procedures	Post a safety notice Perform written safety certification. Provide safety equipment Keep a record of injuries: ■ OSHA Form 300A—Summary of Work-Related Injuries and Illnesses ■ OSHA Form 301—Injuries and Illnesses Incident Report Post Form 300A each February. Report any serious or fatal accidents to the federal or state authority	
Recordkeeping	Keeps records for a sufficient period.	

[YOUR COMPANY]

Family and Medical Leave Response Form

(for requests for leave due to employee's health)

Dear [insert employee's name],

I am writing you to advise you of your rights and obligations when taking leave under the Family and Medical Leave Act of 1993 (the "FMLA"), in accordance with our policy requirements as permitted by that law.

As stated, our policy applies to leaves of absence covered by the FMLA. To be eligible for Family and Medical Leave under the FMLA ("FMLA leave"), you must have been employed by our company for at least 12 months, have worked a minimum of 1,250 hours during the last 12 months, and be employed at a worksite (as that is defined by the law) where 50 or more employees are employed by the company within 75 miles of that worksite. Employees meeting these requirements are permitted to take up to 12 weeks of unpaid leave during any 12-month period for the following family and medical reasons:

- Circumstances involving the birth of a child of an employee or placement with an employee of a child for adoption or foster care;
- For the care of a spouse, child or parent, who has a serious health condition as that term is defined by the FMLA and its regulations; or
- For a serious health condition as defined by the FMLA and its regulations that makes an employee unable to perform his/her job.

Your request for leave due to your health, which you claim makes you unable to perform the functions of your employment position, appears to meet the requirements of our FMLA policy provided you satisfy the medical certification requirements for a serious health condition stated below. Any time you take off from work for this leave will be counted against your FMLA leave entitlement.

To qualify as a "serious health condition" for purposes of the FMLA, an illness, injury, impairment or physical or mental condition must be a present incapacity* that involves one of the following:

- **Hospital Care:** Inpatient care (e.g., an overnight stay) in a hospital, hospice, or residential medical care facility, including any period of incapacity or subsequent treatment in connection with or consequent to such inpatient care; or
- **Absence Plus Treatment:** A period of incapacity requiring absence from work of more than three calendar days (including any subsequent treatment or period of incapacity relating to the same condition) that also involves: (1) treatment two or more times by a health care provider, by a nurse or physician's assistant under direct supervision of a health care provider, or by a provider of health care services (e.g., physical therapist) under orders of, or on referral by, a health care provider; or (2) treatment by a health care provider on at least one occasion which results in a regimen of continuing treatment under the supervision of the health care provider; or
- **Pregnancy:** Any period of incapacity due to pregnancy, or for prenatal care; or
- **Chronic Conditions Requiring Treatments:** A chronic condition: (1) requires periodic visits for treatment by a health care provider, or by a nurse or physician's assistant under direct supervision of a health care provider; (2) continues over an extended period of time (including recurring episodes of a single underlying condition); and (3) may cause episodic rather than a continuing period of incapacity (e.g., asthma, diabetes, epilepsy); or

---

*"Incapacity" for purposes of the FMLA, is defined to mean inability to work, attend school, or perform other regular daily activities due to the serious health condition, treatment therefore, or recovery there from.

- **Permanent/Long-term Conditions Requiring Supervision:** A period of incapacity that is permanent or long-term due to a condition for which treatment may not be effective. The employee must be under the continuing supervision of, but need not be receiving active treatment by a health care provider. Examples include Alzheimer's, a severe stroke, or the terminal stages of a disease; or
- **Multiple Treatments (Non-Chronic Conditions):** Any period of absence to receive multiple treatments (including any period of recovery there from) by a health care provider of health care services under orders of, or on referral by, a health care provider, either for restorative surgery after an accident or other injury, or for a condition that would likely result in a period of incapacity of more than three consecutive calendar days in the absence of medical intervention or treatment, such as cancer (chemotherapy, radiation, etc.), severe arthritis (physical therapy), or kidney disease (dialysis).

Treatment includes examinations to determine if a serious health condition exists and evaluations of the conditions. Treatment does not include routine physical examinations, eye examinations, or dental examinations. A regimen of continuing treatment includes, for example, a course of prescription medication (e.g. an antibiotic) or therapy requiring special equipment to resolve or alleviate the health condition. A regimen of treatment does not include the taking of over-the-counter medications such as aspirin, antihistamines, or salves; or bed rest, drinking fluids, exercise, and other similar activities that can be initiated without a visit to a health care provider. A serious health condition does not include the common cold, influenza, ear aches, upset stomach, minor ulcers, headaches (other than migraines), routine dental, orthodontia problems, or periodontal disease. Further, the serious health condition must make you as an employee unable to perform the essential functions of your position with the company.

In order to establish that you have a "serious health condition" that entitles you to FMLA leave, the company requires that you provide it with a medical certification from your health care provider on the form provided with this letter indicating the following information: the date on which your serious health condition began and the probable duration of that condition, a diagnosis of the condition, a statement of the regimen of treatment prescribed for the condition (including estimated number of visits, nature, frequency and duration of treatment, including treatment by another provider of health services on referral by or order of your health care provider), an indication whether any inpatient hospitalization is required, and a statement that you are unable to perform any work of any kind, or that you are unable to perform the essential functions of your position. This medical certification must be returned to the undersigned within 15 calendar days unless that is not practicable under your particular circumstances and despite your diligent and good faith efforts to obtain the certification within that time limit. Any delay in providing the certification in a timely manner must be justified by you and may result in your leave being denied.

It should be noted that if the company questions the adequacy of the medical certification provided by your health care provider, it may require you to obtain a second opinion at its expense from a health care provider designated by the company to provide that second opinion. If the two opinions differ, the company may require you to obtain a certification from a third health care provider, again at the company's expense, which shall be final and binding. In the event that the company finds it necessary to obtain a third certification, it will discuss with you the selection of the health care provider who will undertake that evaluation.

Any accrued paid vacation, sick leave, or other available paid leave for which you are eligible must be taken at the outset of your FMLA leave before any unpaid leave is to be taken. If such available paid leave is less than twelve weeks, FMLA leave may be used to complete the twelve weeks of FMLA leave.

During your FMLA leave, any health insurance benefits that you were receiving prior to taking FMLA leave will continue on the same basis as if you continued to work. You will be required to continue paying all co-payments of premiums, at the same rate as if you were not on leave. Such co-payments are due by_____. If you wish to make payments in another manner, you must call me to discuss the possibility of any such alternate arrangement. Your health insurance coverage will cease if the premium payments due from you are more than thirty days late or if you wish to discontinue receiving such benefits while on FMLA leave. However, upon returning from FMLA leave, your coverage will be restored. If you do not return to work after your FMLA leave entitlement is exhausted or expires, the company will attempt to recover all health plan premiums it paid for your health insurance coverage during the period of your FMLA leave, unless prohibited by law. If you do not return to work, your health insurance coverage will be terminated and you will be offered continuation coverage under the provisions of COBRA, if you are otherwise eligible for COBRA.

The company requests that you provide it with recertification of your medical condition every 30 days, if the duration of your FMLA leave is in excess of that period, as well as in the following circumstances: where you request an extension of leave (and your initial period of leave was less than twelve weeks in duration), where circumstances described by the original certification have changed significantly, or where the company receives information that casts doubt upon the continuing validity of your original certification. Further, the company will require you to provide a medical certification, when applicable, where you claim that you are unable to return to work after your FMLA leave because of the continuation, recurrence, or onset of a serious health condition which thereby prevents the company from recovering its share of health benefit premium payments made on your behalf during your unpaid FMLA leave.

Since your request for FMLA leave is based upon your having a serious health condition which makes you unable to perform your job, before returning to work you must provide the company with a medical certification from your health care provider that you are able to resume your work. If you fail to provide this requested "fitness-for-duty" certification before your return to work, you may be denied restoration to a position with the company until such certification is submitted.

The company requires that you affirm your intent to return to work at the company by signing and returning a copy of this letter, which is provided, at the space for your signature indicated below. Also, the company may require that while you are on FMLA leave, you report periodically on your status and your intent to, and expected date of, return to active duty with the company.

Upon your timely return to work upon completion of FMLA leave and proper submission of a "fitness-for-duty" certification, you will be returned to your former position or to an equivalent position.

If you have any questions or need clarification on any aspect of your entitlement under our FMLA policy, feel free to contact me at_____.

Sincerely,

Name, Title:_____

☐ I intend to return to work.
☐ I do not intend to return.

Employee Signature:_____ Date:_____

[YOUR COMPANY]

Family and Medical Leave Response Form

(for requests for leave due to the birth of an employee's child and/or the placement of a child with the employee for adoption or foster care)

Dear [insert employee's name],

I am writing you to advise you of your rights and obligations when taking leave under the Family and Medical Leave Act of 1993 (the "FMLA"), in accordance with our policy requirements as permitted by that law.

As stated, our policy applies to leaves of absence covered by the FMLA. To be eligible for Family and Medical Leave under the FMLA ("FMLA leave"), you must have been employed by this company for at least 12 months, have worked a minimum of 1,250 hours during the last 12 months, and be employed at a worksite (as that is defined by the law) where 50 or more employees are employed by the company within 75 miles of that worksite. Employees meeting these requirements are permitted to take up to twelve weeks of unpaid leave during any 12-month period for the following family and medical reasons:

- Circumstances involving the birth of a child of an employee or placement with an employee of a child for adoption or foster care;
- For the care of a spouse, child or parent, who has a serious health condition as that term is defined by the FMLA and its regulations; or
- For a serious health condition as defined by the FMLA and its regulations that makes an employee unable to perform his/her job.

Your request for leave associated with the birth of your child/the placement with you of a child for adoption or foster care meets these requirements and the time you take off from work for this leave will be counted against your FMLA leave entitlement.

Any accrued paid vacation, sick leave, or other available paid leave for which you are eligible must be taken at the outset of your FMLA leave before any unpaid leave is to be taken. If such available paid leave is less than 12 weeks, FMLA leave may be used to complete the 12 weeks of FMLA leave.

During your FMLA leave, any health insurance benefits that you were receiving prior to taking FMLA leave will continue on the same basis as if you continued to work. You will be required to continue paying all co-payments of premiums, at the same rate as if you were not on leave. Such co-payments are due by. If you wish to make payments in another manner, you must call me to discuss the possibility of any such alternate arrangement. Your health insurance coverage will cease if the premium payments due from you are more than 30 days late or if you wish to discontinue receiving such benefits while on FMLA leave. However, upon returning from FMLA leave, your coverage will be restored. If you do not return to work after your FMLA leave entitlement is exhausted or expires, the company will attempt to recover from you all health plan premiums it paid for your health insurance coverage during the period of your FMLA leave, unless prohibited by law. If you do not return to work, your health insurance coverage will be terminated and you will be offered continuation coverage under the provisions of COBRA, if you are otherwise eligible for COBRA.

The company requires that you affirm your intent to return to work at the company by signing and returning a copy of this letter, which is provided, at the space for your signature indicated below. Also, the company may require that while you are on FMLA leave you report periodically on your status and your intent to, and expected date of return to active duty with the company.

Upon your timely return to work upon completion of FMLA leave, you will be returned to your former position or to an equivalent position.

If you have any questions or need clarification on any aspect of your entitlement under our FMLA Policy, feel free to contact me at   .

Sincerely,

Name, Title: _____

☐ I intend to return to work.
☐ I do not intend to return.

Employee Signature: _____ Date: _____

Family and Medical Leave Response Form

(for requests for leave to care for an employee's family member who is ill)

Dear [insert employee's name],

I am writing you to advise you of your rights and obligations when taking leave under the Family and Medical Leave Act of 1993 (the "FMLA"), in accordance with our policy requirements as permitted by that law.

As stated, our policy applies to leaves of absence covered by the FMLA. To be eligible for Family and Medical Leave under the FMLA ("FMLA leave"), you must have been employed by company for at least 12 months, have worked a minimum of 1,250 hours during the last 12 months, and be employed at a worksite (as that is defined by the law) where 50 or more employees are employed by the company within 75 miles of that worksite. Employees meeting these requirements are permitted to take up to twelve weeks of unpaid leave during any 12-month period for the following family and medical reasons:

- Circumstances involving the birth of a child of an employee or placement with an Employee of a child for adoption or foster care;
- For the care of a spouse, child or parent ("family member"), who has a serious health condition as that tern is defined by the FMLA and its regulations; or
- For a serious health condition as defined by the FMLA and its regulations that makes an employee unable to perform his/her job.

Your request for leave to care for your family member who is ill, and whom you claim is incapable of self-care, appears to meet the requirements of our FMLA policy provided you satisfy the medical certification requirements for the care of such a family member with a serious health condition as stated below. Any time you take off from work for this leave will be counted against your FMLA leave entitlement.

To qualify as a "serious health condition" for purposes of the FMLA, an illness, injury, impairment, or physical or mental condition must be a present incapacity* that involves one of the following:

- **Hospital Care:** Inpatient care (e.g., an overnight stay) in a hospital, hospice, or residential medical care facility, including any period of incapacity or subsequent treatment in connection with or consequent to such inpatient care.
- **Absence Plus Treatment:** A period of incapacity requiring absence of more than three calendar days (including any subsequent treatment or period of incapacity relating to the same condition) that also involves: (1) treatment two or more times by a health care provider, by a nurse, or physician's assistant under direct supervision of a health care provider, or by a provider of health care services (e.g., physical therapist) under orders of, or on referral by, a health care provider; or (2) treatment by a health care provider on at least one occasion which results in a regimen of continuing treatment under the supervision of the health care provider.
- **Pregnancy:** Any period of incapacity due to pregnancy, or for prenatal care.
- **Chronic Conditions Requiring Treatments:** A chronic condition: (1) requires periodic visits for treatment by a health care provider, or by a nurse, or physician's assistant under direct supervision of a health care provider; (2) continues over an extended period of time (including recurring episodes of a single underlying condition); and (3) may cause episodic rather than a continuing period of incapacity (e.g., asthma, diabetes, epilepsy).

---

* "Incapacity" for purposes of the FMLA, is defined to mean inability to work, attend school, or perform other regular daily activities due to the serious health condition, treatment therefore, or recovery there from.

- **Permanent/Long-Term Conditions Requiring Supervision:** A period of incapacity that is permanent or long-term due to a condition for which treatment may not be effective. The family member must be under the continuing supervision of, but need not be receiving active treatment by a health care provider. Examples include Alzheimer's, a severe stroke, or the terminal stages of a disease.
- **Multiple Treatments (Non-Chronic Conditions):** Any period of absence to receive multiple treatments (including any period of recovery there from) by a health care provider of health care services under orders of, or on referral by, a health care provider, either for restorative surgery after an accident or other injury, or for a condition that would likely result in a period of incapacity of more than three consecutive calendar days in the absence of medical intervention or treatment, such as cancer (chemotherapy, radiation, etc.), severe arthritis (physical therapy), or kidney disease (dialysis).

Treatment includes examinations to determine if a serious health condition exists and evaluations of the conditions. Treatment does not include routine physical examinations, eye examinations, or dental examinations. A regimen of continuing treatment includes, for example, a course of prescription medication (e.g., an antibiotic) or therapy requiring special equipment to resolve or alleviate the health condition. A regimen of treatment does not include the taking of over-the-counter medications such as aspirin, antihistamines, or salves; or bed rest, drinking fluids, exercise, and other similar activities that can be initiated without a visit to a health care provider. A serious health condition does not include the common cold, influenza, ear aches, upset stomach, minor ulcers, headaches (other than migraines), routine dental, orthodontia problems or periodontal disease.

In order to establish that your family member has a "serious health condition" which entitles you to FMLA Leave, the company requires that you provide it with a medical certification from your family member's health care provider on the form provided with this letter indicating the following information: the patient's name, the date on which the serious health condition began and the probable duration of that condition, a diagnosis of the condition, a statement of the regimen of treatment prescribed for the condition (including estimated number of visits, nature, frequency and duration of treatment, including treatment by another provider of health services on referral by or order of the health care provider), an indication whether any inpatient hospitalization is required, a statement by the health services provider that your presence is necessary or that it would be beneficial for the care of the patient, and a statement by you of the care you will provide and the time period during which the care will be provided. This medical certification must be returned to the undersigned within 15 calendar days unless that is not practicable under your particular circumstances and despite your diligent and good faith efforts to obtain the certification within that time limit. Any delay in providing the certification in a timely manner must be justified by you and may result in your leave being denied.

It should be noted that if the company questions the adequacy of the medical certification provided by the family member's health care provider, it may require you to obtain a second opinion at its expense from a health care provider designated by the company to provide that second opinion. If the two opinions differ, the company may require you to obtain a certification from a third health care provider, again at the company's expense, which shall be final and binding. In the event that the company finds it necessary to obtain a third certification, it will discuss with you the selection of the health care provider who will undertake that evaluation.

Any accrued paid vacation, sick leave or other available paid leave for which you are eligible must be taken at the outset of your FMLA leave before any unpaid leave is to be taken. If such available paid leave is less than twelve weeks, FMLA leave may be used to complete the twelve weeks of FMLA leave.

During your FMLA leave, any health insurance benefits that you were receiving prior to taking FMLA leave will continue on the same basis as if you continued to work. You will be required to continue paying all co-payments of premiums, at the same rate as if you were not on leave. Such co-payments are due by _____. If you wish to make payments in another manner, you must call me to discuss the possibility of any such alternate arrangement. Your health insurance coverage will cease if the premium payments due from you are more than thirty days late or if you wish to discontinue receiving such benefits while on FMLA leave. However, upon returning from FMLA leave, your coverage will be restored. If you do not return to work after your FMLA leave entitlement is exhausted or expires, the company will attempt to recover all health plan premiums it paid the company for your health insurance coverage during the period of your FMLA leave, unless prohibited by law. If you do not return to work, your health insurance coverage will be terminated and you will be offered continuation coverage under the provisions of COBRA, if you are otherwise eligible for COBRA.

The company requests that you provide it with recertification of your family member's medical condition every thirty days, if the duration of your FMLA leave is in excess of that period, as well as in the following circumstances: where you request an extension of leave (and your initial period of leave was less than twelve weeks in duration), where circumstances described by the original certification have changed significantly, or where the company receives information that casts doubt upon the continuing validity of your original certification. Further, the company will require you to provide a medical certification, when applicable, where you claim that you are unable to return to work after your FMLA leave because of the continuation, recurrence, or onset of a serious health condition which thereby prevents the company from recovering its share of health benefit premium payments made on your behalf during your unpaid FMLA leave.

The company requires that you affirm your intent to return to work at the company by signing and returning a copy of this letter, which is provided, at the space for your signature indicated below. Also, the company may require that while you are on FMLA leave, you report periodically on your status and your intent to, and expected date of, return to active duty with the company.

Upon your timely return to work upon completion of FMLA leave, you will be returned to your former position or to an equivalent position.

If you have any questions or need clarification on any aspect of your entitlement under our FMLA Policy, feel free to contact me at _____.

Sincerely,

Name, Title: _____

☐ I intend to return to work.
☐ I do not intend to return.

Employee Signature: _____ Date: _____

Sample Job Description Form

Date Developed: _____ Effective Date: _____

Developer's Name: _____

Position Title: _____

☐ New Position/Revised Position
☐ Exempt/Non Exempt (circle one)

Name and Title of Supervisor: _____

Job Location: _____

Purpose of the Job: (What are the end results of this position's objectives? Why does the job exist?)

Essential Functions: (Describe functions the employee must perform; ones that are fundamental to the position.)

- List functions in sequential or priority order (e.g., the task that requires the most amount of time or carries the greatest responsibility should be listed first.)
- Estimate the percentage of time spent on each function. The amount of time for all duties should add up to 100 percent
- State separate duties clearly and concisely. Be as specific as possible.
- Be objective and accurate. Describe the job as it realistically should be performed.
- To comply with the Americans With Disabilities Act of 1990 (ADA), which prohibits discrimination against qualified individuals on the basis of disability, it is necessary to specify the physical and mental requirements and environmental conditions of the essential functions of the job. Use the form on page 131 to assess the physical and mental demands for this position. Then describe them here.

Non-Essential Functions: (List functions or duties performed occasionally or in addition to the position's essential functions. Those that are peripheral, incidental, or a minimal part of the job are considered non-essential. Here are some tips for identifying them):

- Would removing the duty fundamentally change the job? If not, the duty is non-essential.
- Are there other employees available to perform the duty? If it is feasible to redistribute the work, the duty may be non-essential.

Education/Skills Requirements: (List the minimum required education and skills required to perform this job. Be sure to separate required qualifications from preferred and/or desirable qualifications. You should have the same expectations of skills and education for all employees in the same job title.)

Experience Requirements and/or Equivalents: (List the minimum experience and/or equivalents required to perform this job. As above, be sure to separate required experience from preferred experience. Include licenses and certificates here.)

Supervisory Responsibilities: (Include the number and type of employees supervised and the position's level of authority to hire, fire, make recommendations, assign work, evaluate performance, etc.)

Fiscal Responsibility: (Include budgeting responsibilities, approval privileges on purchase orders and check requests, reporting and auditing functions, etc.)

Extent of Public Contact: (To what degree will this employee interact with members of the public? Include situations inside and outside the company)

Working Conditions and Environment: (Include necessary travel, unusual environmental conditions etc.)

- Use the form on page 134 to assess the working conditions and environment for this position. Then describe them here.

Work Schedule Description: (List typical working hours for this position. Include anything unusual about the work schedule, terms of the employment, worksite changes, etc.)

Add Disclaimers:

- Brief at-will employment statement.
- That the job description is not an express or implied contract.
- That management reserves the right to add, delete, or modify the job description.
- That other duties may be assigned by the supervisor and/or his or her designee.

Physical and Mental Job Demand Assessment Form

Use this form to assess the following physical and mental job demands as they pertain to the essential functions of each position. Indicate how often each essential motion and sensory and mental ability is required. Specify any specific abilities required.

Essential Motion Indicate weight limit in lbs. ( )	N/A	Occasionally 1–33%	Frequently 34–66%	Constantly 67–100%
Pushing ( )				
Pulling ( )				
Lifting ( )				
Carrying ( )				
Sitting				
Standing				
Walking				
Climbing stairs				
Sustained bending				
Overhead reaching				
Crawling				
Squatting (repeated)				
Kneeling				
Stooping (repeated)				
Crouching				
Climbing ladders				
Balancing				
Handling				
Driving				
Digital dexterity (e.g., using computer keyboard, writing manually)				
Essential Sensory Abilities				
Tasting				
Smelling				
Hearing speech				
Hearing mechanical sounds				
Close vision (clear vision at 20 in. or less)				
Distance vision (clear vision at 20 ft. or more)				
Color vision				
Peripheral vision				
Depth perception				
Essential Mental Abilities				
Use written (verbal visual) sources of information (e.g., read reports, manuals, etc.)				
Use non-verbal visual sources of information (e.g., reference graphs, tables, etc.)				
Use oral communication to perform work (e.g., answer telephone, receive visitors, etc.)				
Make minor decisions requiring limited judgment (e.g., task sequencing, sorting mail)				
Make general decisions in the absence of specific directions (e.g., prioritizing)				
Perform activities requiring sustained concentration (e.g., designing, planning, etc.)				

Working Conditions and Environment Assessment Form

Use this form to assess exposure to physical risks and to gather information to describe the working environment for employees in each position. Indicate the frequency with which the employee will be exposed to these conditions.

Environmental Conditions	N/A	Occasionally 1–33%	Frequently 34–66%	Constantly 67–100%
Wet/humid				
Extreme heat (indicate typical temperature(s)				
Extreme cold (indicate typical temperature(s)				
Abrupt temperature change (indicate range)				
Work with or near machinery				
Heights				
Work with explosives				
Vibration				
Risk of radiation				
Risk of electric shock				
Noise Levels				
Very quiet (faintest sounds heard by ear)				
Quiet (leaves rustling, library)				
Moderate (light traffic, business office)				
Loud (manufacturing plant, lawnmower)				
Very loud (jack hammer, jet engine)				
Atmospheric Conditions				
Fumes				
Mist				
Odors				
Gasses				
Dust				
Poor ventilation				

Form **SS-8** (Rev. January 2001) Department of the Treasury Internal Revenue Service	**Determination of Worker Status for Purposes of Federal Employment Taxes and Income Tax Withholding**	OMB No. 1545-0004

Name of firm (or person) for whom the worker performed services	Worker's name
Firm's address (include street address, apt. or suite no., city, state, and ZIP code)	Worker's address (include street address, apt. or suite no., city, state, and ZIP code)

Trade name	Telephone number (include area code) ( )	Worker's social security number

Telephone number (include area code) ( )	Firm's employer identification number	Worker's employer identification number (if any)

Important Information Needed To Process Your Request

If this form is being completed by the worker, the IRS must have your permission to disclose your name to the firm. Do you object to disclosing your name and the information on this form to the firm? . . . . ☐ **Yes** ☐ **No**
If you answered "Yes" or did not check a box, stop here. The IRS cannot act on your request and a determination will not be issued.

You must answer ALL items OR mark them "Unknown" or "Does not apply." If you need more space, attach another sheet.

A This form is being completed by: ☐ Firm ☐ Worker; for services performed _____ to _____ .
(beginning date) (ending date)

B Explain your reason(s) for filing this form (e.g., you received a bill from the IRS, you believe you received a Form 1099 or Form W-2 erroneously, you are unable to get worker's compensation benefits, you were audited or are being audited by the IRS). ------------------------------
----

C Total number of workers who performed or are performing the same or similar services _____ .

D How did the worker obtain the job? ☐ Application ☐ Bid ☐ Employment Agency ☐ Other (specify) _____

E Attach copies of all supporting documentation (contracts, invoices, memos, Forms W-2, Forms 1099, IRS closing agreements, IRS rulings, etc.). In addition, please inform us of any current or past litigation concerning the worker's status. If no income reporting forms (Form 1099-MISC or W-2) were furnished to the worker, enter the amount of income earned for the year(s) at issue $ _____ .

F Describe the firm's business. --

G Describe the work done by the worker and provide the worker's job title. --------------------------
----

H Explain why you believe the worker is an employee or an independent contractor. -----------------
----

I Did the worker perform services for the firm before getting this position? . . . . ☐ **Yes** ☐ **No** ☐ **N/A**
If "Yes," what were the dates of the prior service? --
If "Yes," explain the differences, if any, between the current and prior service. --------------------
----

J If the work is done under a written agreement between the firm and the worker, attach a copy (preferably signed by both parties). Describe the terms and conditions of the work arrangement. --------------------------
----

For Privacy Act and Paperwork Reduction Act Notice, see page 5. Cat. No. 16106T Form **SS-8** (Rev. 1-2001)

Part I **Behavioral Control**

1 What specific training and/or instruction is the worker given by the firm? ....

....

2 How does the worker receive work assignments? ....

....

3 Who determines the methods by which the assignments are performed? ....

4 Who is the worker required to contact if problems or complaints arise and who is responsible for their resolution? ....

....

5 What types of reports are required from the worker? Attach examples. ....

....

6 Describe the worker's daily routine (i.e., schedule, hours, etc.). ....

....

....

7 At what location(s) does the worker perform services (e.g., firm's premises, own shop or office, home, customer's location, etc.)? ....

....

8 Describe any meetings the worker is required to attend and any penalties for not attending (e.g., sales meetings, monthly meetings, staff meetings, etc.). ....

9 Is the worker required to provide the services personally? . . . . ☐ **Yes** ☐ **No**

10 If substitutes or helpers are needed, who hires them? ....

11 If the worker hires the substitutes or helpers, is approval required? . . . . ☐ **Yes** ☐ **No**

If "Yes," by whom? ....

12 Who pays the substitutes or helpers? ....

13 Is the worker reimbursed if the worker pays the substitutes or helpers? . . . . ☐ **Yes** ☐ **No**

If "Yes," by whom? ....

Part II **Financial Control**

1 List the supplies, equipment, materials, and property provided by each party:

The firm ....

The worker ....

Other party ....

2 Does the worker lease equipment? . . . . ☐ **Yes** ☐ **No**

If "Yes," what are the terms of the lease? (Attach a copy or explanatory statement.) ....

....

3 What expenses are incurred by the worker in the performance of services for the firm? ....

....

4 Specify which, if any, expenses are reimbursed by:

The firm ....

Other party ....

5 Type of pay the worker receives: ☐ Salary   ☐ Commission   ☐ Hourly Wage   ☐ Piece Work

☐ Lump Sum   ☐ Other (specify) ....

If type of pay is commission, and the firm guarantees a minimum amount of pay, specify amount $ _____ .

6 If the worker is paid by a firm other than the one listed on this form for these services, enter name, address, and employer identification number of the payer. ....

....

7 Is the worker allowed a drawing account for advances? . . . . ☐ **Yes** ☐ **No**

If "Yes," how often? ....

Specify any restrictions. ....

....

8 Whom does the customer pay? . . . . ☐ Firm   ☐ Worker

If worker, does the worker pay the total amount to the firm?   ☐ **Yes** ☐ **No** If "No," explain. ....

....

9 Does the firm carry worker's compensation insurance on the worker? . . . . ☐ **Yes** ☐ **No**

10 What economic loss or financial risk, if any, can the worker incur beyond the normal loss of salary (e.g., loss or damage of equipment, material, etc.)? ....

....

Form **SS-8** (Rev. 1-2001)

Part III   **Relationship of the Worker and Firm**

1   List the benefits available to the worker (e.g., paid vacations, sick pay, pensions, bonuses). ....

2   Can the relationship be terminated by either party without incurring liability or penalty? . . . . ☐ **Yes**   ☐ **No**
If "No," explain your answer. ....

3   Does the worker perform similar services for others? . . . . ☐ **Yes**   ☐ **No**
If "Yes," is the worker required to get approval from the firm? . . . . ☐ **Yes**   ☐ **No**

4   Describe any agreements prohibiting competition between the worker and the firm while the worker is performing services or during any later period. Attach any available documentation. ....

5   Is the worker a member of a union? . . . . ☐ **Yes**   ☐ **No**

6   What type of advertising, if any, does the worker do (e.g., a business listing in a directory, business cards, etc.)? Provide copies, if applicable.
....

7   If the worker assembles or processes a product at home, who provides the materials and instructions or pattern? ....

8   What does the worker do with the finished product (e.g., return it to the firm, provide it to another party, or sell it)? ....

9   How does the firm represent the worker to its customers (e.g., employee, partner, representative, or contractor)? ....

10 If the worker no longer performs services for the firm, how did the relationship end? ....

Part IV   **For Service Providers or Salespersons**—Complete this part if the worker provided a service directly to customers or is a salesperson.

1   What are the worker's responsibilities in soliciting new customers? ....

2   Who provides the worker with leads to prospective customers? ....

3   Describe any reporting requirements pertaining to the leads. ....

4   What terms and conditions of sale, if any, are required by the firm? ....

5   Are orders submitted to and subject to approval by the firm? . . . . ☐ **Yes**   ☐ **No**

6   Who determines the worker's territory? ....

7   Did the worker pay for the privilege of serving customers on the route or in the territory? . . . . ☐ **Yes**   ☐ **No**
If "Yes," whom did the worker pay? ....
If "Yes," how much did the worker pay? . . . . $ _____

8   Where does the worker sell the product (e.g., in a home, retail establishment, etc.)? ....

9   List the product and/or services distributed by the worker (e.g., meat, vegetables, fruit, bakery products, beverages, or laundry or dry cleaning services). If more than one type of product and/or service is distributed, specify the principal one. ....

10 Does the worker sell life insurance full time? . . . . ☐ **Yes**   ☐ **No**

11 Does the worker sell other types of insurance for the firm? . . . . ☐ **Yes**   ☐ **No**
If "Yes," enter the percentage of the worker's total working time spent in selling other types of insurance. . . . _____ %

12 If the worker solicits orders from wholesalers, retailers, contractors, or operators of hotels, restaurants, or other similar establishments, enter the percentage of the worker's time spent in the solicitation. . . . . _____ %

13 Is the merchandise purchased by the customers for resale or use in their business operations? . . . . ☐ **Yes**   ☐ **No**
Describe the merchandise and state whether it is equipment installed on the customers' premises. ....

Part V   **Signature** (see page 4)

Under penalties of perjury, I declare that I have examined this request, including accompanying documents, and to the best of my knowledge and belief, the facts presented are true, correct, and complete.

Signature ▶ _____   Title ▶ _____   Date ▶ _____
(Type or print name below)

Part III Relationship of the Worker and Firm

1 List the benefits available to the worker (e.g., paid vacations, sick pay, pensions, bonuses). ....

2 Can the relationship be terminated by either party without incurring liability or penalty? . . . . ☐ Yes ☐ No
If "No," explain your answer. ....

3 Does the worker perform similar services for others? . . . . ☐ Yes ☐ No
If "Yes," is the worker required to get approval from the firm? . . . . ☐ Yes ☐ No

4 Describe any agreements prohibiting competition between the worker and the firm while the worker is performing services or during any later period. Attach any available documentation. ....

5 Is the worker a member of a union? . . . . ☐ Yes ☐ No

6 What type of advertising, if any, does the worker do (e.g., a business listing in a directory, business cards, etc.)? Provide copies, if applicable. ....

7 If the worker assembles or processes a product at home, who provides the materials and instructions or pattern? ....

8 What does the worker do with the finished product (e.g., return it to the firm, provide it to another party, or sell it)? ....

9 How does the firm represent the worker to its customers (e.g., employee, partner, representative, or contractor)? ....

10 If the worker no longer performs services for the firm, how did the relationship end? ....

Part IV For Service Providers or Salespersons—Complete this part if the worker provided a service directly to customers or is a salesperson.

1 What are the worker's responsibilities in soliciting new customers? ....

2 Who provides the worker with leads to prospective customers? ....

3 Describe any reporting requirements pertaining to the leads. ....

4 What terms and conditions of sale, if any, are required by the firm? ....

5 Are orders submitted to and subject to approval by the firm? . . . . ☐ Yes ☐ No

6 Who determines the worker's territory? ....

7 Did the worker pay for the privilege of serving customers on the route or in the territory? . . . . ☐ Yes ☐ No
If "Yes," whom did the worker pay? ....
If "Yes," how much did the worker pay? . . . . $ _____

8 Where does the worker sell the product (e.g., in a home, retail establishment, etc.)? ....

9 List the product and/or services distributed by the worker (e.g., meat, vegetables, fruit, bakery products, beverages, or laundry or dry cleaning services). If more than one type of product and/or service is distributed, specify the principal one. ....

10 Does the worker sell life insurance full time? . . . . ☐ Yes ☐ No

11 Does the worker sell other types of insurance for the firm? . . . . ☐ Yes ☐ No
If "Yes," enter the percentage of the worker's total working time spent in selling other types of insurance. . . . _____ %

12 If the worker solicits orders from wholesalers, retailers, contractors, or operators of hotels, restaurants, or other similar establishments, enter the percentage of the worker's time spent in the solicitation. . . . . _____ %

13 Is the merchandise purchased by the customers for resale or use in their business operations? . . . . ☐ Yes ☐ No
Describe the merchandise and state whether it is equipment installed on the customers' premises. ....

Part V Signature (see page 4)

Under penalties of perjury, I declare that I have examined this request, including accompanying documents, and to the best of my knowledge and belief, the facts presented are true, correct, and complete.

Signature ▶ _____ Title ▶ _____ Date ▶ _____
(Type or print name below)

General Instructions

Section references are to the Internal Revenue Code unless otherwise noted.

Purpose

Firms and workers file Form SS-8 to request a determination of the status of a worker for purposes of Federal employment taxes and income tax withholding.

A Form SS-8 determination may be requested only in order to resolve Federal tax matters. The taxpayer requesting a determination must file an income tax return for the years under consideration before a determination can be issued. If Form SS-8 is submitted for a tax year for which the statute of limitations on the tax return has expired, a determination letter will not be issued. The statute of limitations expires 3 years from the due date of the tax return or the date filed, whichever is later.

The IRS does not issue a determination letter for proposed transactions or on hypothetical situations. We may, however, issue an information letter when it is considered appropriate.

Definition

Firm. For the purposes of this form, the term "firm" means any individual, business enterprise, organization, state, or other entity for which a worker has performed services. The firm may or may not have paid the worker directly for these services. **If the firm was not responsible for payment for services, please be sure to complete question 6 in Part II of Form SS-8.**

The SS-8 Determination Process

The IRS will acknowledge the receipt of your Form SS-8. Because there are usually two (or more) parties who could be affected by a determination of employment status, the IRS attempts to get information from all parties involved by sending those parties blank Forms SS-8 for completion. The case will be assigned to a technician who will review the facts, apply the law, and render a decision. The technician may ask for additional information before rendering a decision. The IRS will generally issue a formal determination to the firm or payer (if that is a different entity), and will send a copy to the worker. A determination letter applies only to a worker (or a class of workers) requesting it, and the decision is binding on the IRS. In certain cases, a formal determination will not be issued; instead, an information letter may be issued. Although an information letter is advisory only and is not binding on the IRS, it may be used to assist the worker to fulfill his or her Federal tax obligations. This process takes approximately 120 days.

Neither the SS-8 determination process nor the review of any records in connection with the determination constitutes an examination (audit) of any Federal tax return. If the periods under consideration have previously been examined, the SS-8 determination process will not constitute a reexamination under IRS reopening procedures. Because this is not an examination of any Federal tax return, the appeal rights available in connection with an examination do not apply to an SS-8 determination. However, if you disagree with a determination and you have additional information concerning the work relationship that you believe was not previously considered, you may request that the determining office reconsider the determination.

Completing Form SS-8

Please answer all questions as completely as possible. Attach additional sheets if you need more space. Provide information for all years the worker provided services for the firm. Determinations are based on the entire relationship between the firm and the worker.

Additional copies of this form may be obtained by calling 1-800-TAX-FORM (1-800-829-3676) or from the IRS Web Site at **www.irs.gov.**

Fee

There is no fee for requesting an SS-8 determination letter.

Signature

The Form SS-8 must be signed and dated by the taxpayer. A stamped signature will not be accepted.

The person who signs for a corporation must be an officer of the corporation who has personal knowledge of the facts. If the corporation is a member of an affiliated group filing a consolidated return, it must be signed by an officer of the common parent of the group.

The person signing for a trust, partnership, or limited liability company must be, respectively, a trustee, general partner, or member-manager who has personal knowledge of the facts.

Where To File

Send the completed Form SS-8 to the address listed below for the firm's location. However, for cases involving Federal agencies, send the form to the Internal Revenue Service, Attn: CC:CORP:T:C, Ben Franklin Station, P.O. Box 7604, Washington, DC 20044.

Firm's location:	Send to:
Alaska, Arizona, Arkansas, California, Colorado, Hawaii, Idaho, Illinois, Iowa, Kansas, Minnesota, Missouri, Montana, Nebraska, Nevada, New Mexico, North Dakota, Oklahoma, Oregon, South Dakota, Texas, Utah, Washington, Wisconsin, Wyoming, American Samoa, Guam, Puerto Rico, U.S. Virgin Islands	Internal Revenue Service SS-8 Determinations P.O. Box 1231 Stop 4106 AUCSC Austin, TX 78767
Alabama, Connecticut, Delaware, District of Columbia, Florida, Georgia, Indiana, Kentucky, Louisiana, Maine, Maryland, Massachusetts, Michigan, Mississippi, New Hampshire, New Jersey, New York, North Carolina, Ohio, Pennsylvania, Rhode Island, South Carolina, Tennessee, Vermont, Virginia, West Virginia, all other locations not listed	Internal Revenue Service SS-8 Determinations 40 Lakemont Road Newport, VT 05855-1555

Instructions for Workers

If you are requesting a determination for more than one firm, complete a separate Form SS-8 for each firm.

 Form SS-8 is not a claim for refund of social security and Medicare taxes or Federal income tax withholding.

If you are found to be an employee, you are responsible for filing an amended return for any corrections related to this decision. A determination that a worker is an employee does not necessarily reduce any current or prior tax liability. For more information, call 1-800-829-1040.

Forma **SS-8PR** (Rev. enero de 2001) Department of the Treasury Internal Revenue Service	**DETERMINACION DEL ESTADO DE EMPLEO DE UN TRABAJADOR PARA PROPOSITOS DE LAS CONTRIBUCIONES FEDERALES SOBRE EL EMPLEO**	OMB No. 1545-0004

Nombre de la empresa (o individuo) para quien el trabajador realizó los servicios	Nombre del trabajador

Dirección de la empresa (incluya la dirección completa, no. de apto., ciudad o pueblo y zona postal)	Dirección del trabajador (incluya la dirección completa, no. de apto., ciudad o pueblo y zona postal)

Nombre comercial (o nombre usado al operar su negocio)	Número de teléfono (incluya zona telefónica) ( )	Número de seguro social del trabajador

Número de teléfono (incluya zona telefónica) ( )	Número de identificación patronal de la empresa	Número de identificación del patrono del trabajador (si es aplicable)

Información importante que se necesita para procesar su solicitud

Si el trabajador está llenando esta forma, el *IRS* debe tener su permiso para divulgar su nombre a la empresa.

¿Se opone usted a divulgar su nombre y la información contenida en esta forma a la empresa? . . . . . . ☐ **Sí** ☐ **No**

Si usted contestó **"Sí"**, o si no marcó ningún encasillado, el *IRS* no puede tomar ninguna acción referente a su solicitud y no se le emitirá una determinación de su estado de empleo.

Debe responder a TODAS las partidas O marcarlas "Desconozco" o "No aplica". Si necesita más espacio, adjunte otra hoja.

A Está completando esta forma: ☐ La empresa   ☐ El trabajador   por servicios prestados del _____ al _____

B Explique el (los) motivo(s) por el (los) cual(es) usted radica esta planilla (p.e., recibió una factura del *IRS*, recibió una Forma 1099 erróneamente, no puede obtener el derecho de recibir beneficios de la compensación del seguro obrero, sufre o ha sufrido una inspección (auditoría) del *IRS*): ....

....

....

....

C Número total de trabajadores que han prestado o están prestando los mismos o semejantes servicios: _____

D ¿Cómo obtuvo el empleo el trabajador?: ☐ Solicitud   ☐ Propuesta   ☐ De palabra
☐ Agencia de empleos   ☐ Otro ....

E Por favor, adjunte copias de toda clase de documentación que compruebe su estado de empleo (contratos, facturas, memoranda, Formas 499 R-2/W-2 PR, Formas 1099, acuerdos de cierre de auditoría del *IRS,* decisiones o fallos dictados por el *IRS,* etc.). Además, por favor, estado de empleo del trabajador. Si el trabajador no recibió ninguna planilla de reportación de ingresos (Forma 1099-MISC o Forma 499 R-2/W-2 PR), anote la cantidad de ingresos ganados para el (los) año(s) en cuestión: $ _____

F Describa la ocupación principal de la empresa: ....

....

....

....

G Describa los deberes o tareas del trabajador y el título que éste lleva: ....

....

....

....

H Explique por qué usted cree que el trabajador es empleado o es contratista independiente: ....

....

....

....

I ¿Trabajó usted para la empresa antes de aceptar su puesto actual con la misma? . . ☐ **Sí** ☐ **No** ☐ **No aplica**
Si marcó **"Sí"**, anote las fechas de su servicio en esa posición.
Si marcó **"Sí"**, explique las diferencias en los servicios provistos, si alguna, entre su antiguo puesto y el que tiene ahora: ....

....

....

J Si el trabajo se hace bajo un acuerdo por escrito entre la empresa y el trabajador, adjunte una copia (firmada por los dos contratantes, si es posible). Describa los términos y condiciones del arreglo de trabajo: ....

Para el aviso sobre la Ley de Confidencialidad de Información y la Ley de Reducción de Trámites, vea las instrucciones en la página 5.   Cat. No. 23365E   Forma **SS-8PR** (Rev. 1-2001)

| **Parte I** | **Control de las Funciones del Trabajador** |

1 ¿Qué adiestramiento y/o instrucciones específicas recibe el trabajador de la firma? ....
....

2 ¿Cómo recibe el trabajador sus tareas relacionadas con el trabajo? ....

3 ¿Quién determina los métodos que usa el trabajador al desempeñar sus tareas? ....

4 ¿Con quién se pone en contacto el trabajador si surgen quejas o problemas y quién es responsable de resolverlos? ....
....

5 ¿Qué clases de informes o reportes debe someter el trabajador? Incluya unos ejemplares. ....

6 Describa la práctica diaria del trabajador (p.e., horario normal, horas de trabajo, etc.): ....
....

7 ¿En qué localidad (o localidades) presta los servicios el trabajador (p.e., establecimiento de la empresa, su propia oficina o tienda, residencia, establecimiento del cliente, etc.)? ....

8 Describa cualesquier reuniones o conferencias a que el trabajador debe asistir y las penalidades por no asistir (p.e., reuniones de vendedores, reuniones mensuales, reuniones del personal, etc.): ....

9 ¿Se requiere que el trabajador preste los servicios en persona? . . . . ☐ **Sí** ☐ **No**

10 Si se necesita un substituto o ayudante, ¿quién lo contrata? ....

11 Si el trabajador contrata al substituto o ayudante, ¿está éste sujeto a aprobación? . . . . ☐ **Sí** ☐ **No**
Si marcó **"Sí"**, ¿de quién? ....

12 ¿Quién paga al substituto o ayudante? ....

13 ¿Recibe un reembolso el trabajador si éste paga al substituto o ayudante? . . . . ☐ **Sí** ☐ **No**
Si marcó **"Sí"**, ¿de quién? ....

| **Parte II** | **Control Financiero** |

1 Anote aquí las herramientas, equipo, suministros y otros materiales provistos por cada parte:
La empresa ....
El trabajador ....
Otra parte ....

2 ¿Arrienda el trabajador equipo?   ☐ **Sí**   ☐ **No**   Si marcó **"Sí"**, ¿cuáles son las condiciones del contrato de arriendo? (Adjunte copia o una explicación): ....
....

3 ¿Cuáles gastos sufre el trabajador al desempeñar sus servicios por la empresa? ....
....

4 Especifique cuáles son los gastos (si los hay) reembolsados al trabajador por:
La empresa ....
Otra parte ....

5 Tipo de paga que recibe el trabajador: ☐ Sueldo   ☐ Comisiones   ☐ Salario por hora
☐ Salario por ajuste (destajo)   ☐ En suma global   ☐ Otro (especifique) ....
Si su tipo de paga es a base de comisiones y la empresa garantiza una cantidad mínima de remuneración, especifique dicha cantidad: $ _____

6 Si el trabajador recibe su paga de una empresa que no sea la que aparece en esta planilla, anote aquí el nombre, dirección y *EIN* de la misma: ....
....

7 ¿Permite la empresa al trabajador una cuenta de adelantos? . . . . ☐ **Sí** ☐ **No**
Si marcó **"Sí"**, ¿con qué frecuencia puede usarla? ....
Especifique cualesquier restricciones: ....
....

8 ¿A quién paga el cliente? . . . . ☐ A la empresa   ☐ Al trabajador
Si es el trabajador, ¿remite éste la cantidad total que recibe a la empresa? ☐ **Sí**   ☐ **No** Si marcó **"No"**, explique la razón: ....
....

9 ¿Paga la empresa compensación del seguro obrero para el trabajador? . . . . ☐ **Sí** ☐ **No**

10 ¿Qué tipo de pérdida económica o riesgo financiero puede sufrir el trabajador además de la pérdida usual de su sueldo o salario (p.e., pérdida, averío, daño a su equipo, herramientas, materiales, etc.)? . . . .
....

Parte III   Relación entre las Partes

1   Anote los beneficios a la disposición del trabajador (p.e., vacaciones pagadas, compensación por enfermedad, pensiones, bonificaciones, etc.): ....

2   ¿Puede terminar la relación cualquiera de las dos partes sin incurrir en una responsabilidad o penalidad?   ☐ Sí   ☐ No
    Si marcó **"No"**, explique por qué: ....

3   ¿Realiza el trabajador servicios similares para otros? . . . .   ☐ Sí   ☐ No
    Si marcó **"Sí"**, ¿debe el trabajador obtener la empresa primero su aprobación? . . . .   ☐ Sí   ☐ No

4   Describa el acuerdo (si lo hay) entre el trabajador y la empresa que prohibe al trabajador competir con la empresa durante su período de servicio con la empresa e inmediatamente después. Incluya cualquier documentación comprobante: ....

5   ¿Pertenece el trabajador a un sindicato o unión? . . . .   ☐ Sí   ☐ No

6   ¿Qué tipo de publicidad (si alguna) hace el trabajador (p.e., anuncio en el directorio de negocios, tarjeta de representación)? Por favor, provea copias si le corresponde: ....

7   Si el trabajador monta, ensambla o procesa un producto en casa, ¿quién provee los materiales, instrucciones y/o modelos? ....

8   Indique cómo se trata el producto final (p.e., se lo devuelve a la empresa, se lo provee a otra entidad o lo vende): ....

9   Indique cómo representa la empresa al trabajador ante sus clientes (p.e., como empleado, socio, representante o contratista): ....

10 Si el trabajador ya no desempeña servicios para quien le pagaba, ¿cómo se acabó la relación entre los dos? ....

Parte IV   Para los Proveedores de Servicios o los Vendedores—Complete esta parte si el trabajador presta servicios directamente a los clientes o es un vendedor

1   ¿Tiene el trabajador alguna responsabilidad en la obtención de clientes nuevos? ....

2   ¿Quién provee información al trabajador sobre clientes futuros? ....

3   Describa cualesquier requisitos de reportación sobre clientes futuros: ....

4   ¿Cuáles son los términos y condiciones de venta (si los hay) establecidos por la empresa? ....

5   ¿Hay que someter los pedidos o encargos a la empresa para que ésta los apruebe? . . . .   ☐ Sí   ☐ No

6   ¿Quién determina el territorio específico del trabajador? ....

7   ¿Le pagó el trabajador a la empresa o a la persona por el privilegio de servir a los clientes en la ruta o territorio?   ☐ Sí   ☐ No
    Si marcó **"Sí"**, ¿a quién le pagó el trabajador? ....
    Si marcó **"Sí"**, ¿cuánto le pagó al cliente el trabajador? . . . . $ _____

8   ¿Dónde vende el trabajador su producto (p.e., en una residencia, negocio al detal, etc.)? ....

9   Enumere los productos y/o servicios distribuidos por el trabajador, tales como carnes, legumbres, frutas, productos de harina (pan, pasteles, bizcochos, etc.), bebidas (que no sean leche), o servicios de ropa lavada en máquina o en seco. Si se distribuye más de un tipo de producto, especifique el más importante: ....

10 ¿Vende el trabajador seguros de vida todo el tiempo? . . . .   ☐ Sí   ☐ No

11 ¿Vende el trabajador otro(s) tipo(s) de seguros para la empresa? . . . .   ☐ Sí   ☐ No
    Si contestó **"Sí"**, indique el porcentaje de sus horas de trabajo que el trabajador pasó en vender esos otros tipos de seguros . . . .   _____ %

12 Si el trabajador solicita pedidos de mayoristas, detallistas, contratistas u operadores de hoteles, restaurantes o cualquier otro establecimiento similar, especifique el porcentaje de tiempo que el trabajador pasó solicitando pedidos . . . .   _____ %

13 ¿Compran las mercancías los clientes para reventa o para uso en sus operaciones comerciales? . . .   ☐ Sí   ☐ No
    Describa las mercancías e indique si se trata de equipo instalado en el local de negocio de los clientes: ....

Parte V   Firma

Bajo pena de perjurio, declaro que he examinado esta solicitud, incluyendo cualesquier documentos adjuntos y, según mi mejor saber y conocer, los hechos presentados aquí son verídicos, correctos y completos.

Firma ▶ _____   Título ▶ _____   Fecha ▶ _____

Escriba a maquinilla o con letras de molde el nombre del individuo que firmó arriba ....

Instrucciones Generales

Aviso: Las secciones citadas en estas instrucciones se refieren al Código Federal de Rentas Internas, a menos que se indique de otra manera.

Propósito de esta Forma

Las empresas y los trabajadores radican la Forma SS-8PR para obtener una determinación sobre si un trabajador es un empleado para propósitos de las contribuciones federales sobre el empleo.

Se puede solicitar la Forma SS-8PR únicamente para resolver cuestiones relacionadas con las contribuciones federales. El contribuyente que solicita una determinación deberá haber radicado una planilla de contribución para los años en cuestión antes de que se le pueda permitir una determinación. Si se somete una Forma SS-8PR para un año contributivo por el cual se ha vencido la ley de prescripción, no se le emitirá una carta de determinación. La ley de prescripción vence 3 años después de la fecha de vencimiento para radicar la planilla o de la fecha en la cual se radicó, lo que ocurra por último.

El Servicio Federal de Rentas Internas (*IRS*) no emite cartas de determinación sobre asuntos propuestos ni sobre situaciones hipotéticas. Sin embargo, podemos emitir una carta informativa cuando lo consideramos apropiado.

Definición

Empresa. A todos los efectos de esta forma, la palabra "empresa" significa un individuo, empresa comercial, organización, Estado u otra entidad por la cual un trabajador haya prestado servicios. La empresa quizás pagó o no tuvo que pagarle directamente al trabajador por tales servicios. *Si la empresa no fue responsable de pagarle al trabajador por sus servicios, por favor, anote el nombre del pagador en la pregunta 6 de la Parte II de la Forma SS-8PR.*

Los Procedimientos de la Determinación de una Forma SS-8PR

El *IRS* tiene que acusar recibo de su Forma SS-8PR. Ya que normalmente hay dos (o más) partes que pueden ser afectadas por una determinacion del estado de un trabajador, el *IRS* trata de solicitarles información a todas las partes en cuestión mediante emitirles unas Formas SS-8PR en blanco que deberán ser completadas por dichas partes. Su caso será asignado a un experto técnico y éste examinará todos los hechos, aplicará la ley al caso y llegará a una decisión definitiva. El experto pudiera solicitarle a usted más información antes de rendirle una decisión. Por regla general, el *IRS* le emitirá una determinación formal a la empresa o pagador (si éste es distinto a la empresa); una copia de la misma será enviada al trabajador. Una carta de determinación corresponde sólo al trabajador (o clase de trabajadores) que la solicita y el *IRS* está obligado a aceptarla. En algunos casos, no se le emitirá una determinación formal; en vez de eso, se le emitirá una carta informativa. Aunque la carta informativa es de naturaleza consultiva y no le obliga al *IRS* aceptarla formalmente, se puede utilizar para ayudarle al trabajador cumplir con sus deberes contributivos federales. Este proceso suele durar aproximadamente 120 días para resolverse.

Ni los procedimientos para solicitar una determinación en la Forma SS-8PR ni la inspección de cualesquier documentos relacionados con tal determinación constituyen una revisión de una planilla para la declaración de contribución federal. Si los períodos contributivos en cuestión han sido revisados, no se considerará una reconsideración de una revisión previamente hecha de acuerdo con los procedimientos del *IRS* para reconsiderar un caso. Una determinación a base de la Forma SS-8PR emitida a petición de un pagador o trabajador no se ha otorgado con respecto a una revisión y, por consiguiente, los derechos de apelación relacionados con una revisión no entran en el caso. Si no concuerda con los resultados y si tiene más información referente a la relación laborable que existe entre usted y el trabajador y este detalle no se consideró anteriormente, usted puede pedir que la oficina que hizo la determinación original reconsidere el caso.

Cómo Completar la Forma SS-8PR

Conteste todas las preguntas lo más completamente posible. Adjunte hojas adicionales si necesita más espacio. Por favor, incluya en la Forma SS-8PR todos los años durante los cuales el trabajador prestó servicios para la empresa. Se basan las determinaciones en la totalidad de la relación que existe entre la empresa y el trabajador.

Se pueden obtener copias adicionales de esta planilla llamando al 1-800-TAX-FORM (1-800-829-3676) o del sitio que el *IRS* tiene en la red: *www.irs.gov.*

Cargos

No hay que pagar por solicitar una carta de determinación en la Forma SS-8PR.

Requisito de Firma

El contribuyente debe firmar y fechar la Forma SS-8PR. No se permite usar una firma estampada.

La persona que firma por una corporación o sociedad tiene que ser un oficial o ejecutivo de la misma. Esta persona debe tener un conocimiento profundo de los hechos. Si la corporación es socio de un conjunto de negocios afiliados que radican planillas consolidadas, la declaración de pena por perjurio tiene que ser firmada y sometida por un oficial o ejecutivo de la corporación o sociedad matriz del conjunto.

La persona que firma por un fideicomiso, sociedad o corporación de responsabilidad contributiva limitada tiene que ser, respectivamente, un fiduciario, socio general, o socio-gerente que tiene conocimiento personal de todos los hechos.

Adónde se Radica

Para las empresas localizadas en Puerto Rico (y en las Islas Vírgenes), por favor, envíe la Forma SS-8PR al:

Internal Revenue Service
SS-8 Determinations
P.O. Box 1231
Stop 4106 AUCSC
Austin, TX 78767

Instrucciones para los Trabajadores

Si solicita una determinación para más de una sola empresa, llene una Forma SS-8PR por separado para cada empresa en cuestión.

Cuidado: La Forma SS-8PR **no** es una reclamación para un reembolso de las contribuciones al seguro social y al seguro Medicare.

Si se determina que usted es un empleado, usted mismo será resonable de radicar una planilla enmendada para hacer cualesquier correcciones relativas a esta determinación. Además, una determinación de que un individuo es un empleado no necesariamente reducirá una deuda contributiva actual o previa. Llame al 1-800-829-1040 para más información.

Plazo para radicar una reclamación de reembolso. Por regla general, usted debe radicar una reclamación de crédito o de reembolso dentro de 3 años a partir de la fecha en que radicó su planilla original o dentro de 2 años a partir de la fecha en que pagó la contribución, lo que ocurra por último.

La radicación de una Forma SS-8PR no constituye la radicación de una reclamación precautoria, ni tampoco impide el vencimiento del plazo durante el cual se debe radicar una reclamación de reembolso. Si le preocupa el estado de su reembolso y la ley de prescripción para radicar una reclamación de reembolso para el año en cuestión todavía no se ha vencido, usted deberá radicar una reclamación precautoria en la **Forma 1040X,** Planilla Enmendada para la Declaración de la Contribución, o en la **Forma 843,** Reclamación de Reembolso o Solicitud de Reducción de la Contribución, ambas en inglés.

Radique una Forma 1040X por separado para cada año en cuestión. En la Forma 1040X, por favor, deje las líneas de la **1** a la **24** en blanco. Escriba *"Protective Claim* (Reclamación Precautoria)" en la parte superior de la forma, fírmela y féchela. Además, usted deberá anotar las palabras siguientes en la Parte II, *Explanation of Income, Deductions, and Credits* (Explicación de Ingresos, Deducciones y Créditos): *"Filed Form SS-8PR with the IRS office in* (Radiqué la Forma SS-8PR ante la oficina del *IRS* en) *San Juan, PR. By filing this protective claim, I reserve the right to file a claim for any refund that may be due after a determination of my employment tax status has been completed* (Al radicar esta reclamación precautoria, reservo el derecho de radicar una reclamación de cualquier reembolso que pueda resultar después de que se haya llegado a una determinación de mi estado de trabajador con respecto a las contribuciones por razón del empleo.)."

La radicación de una Forma SS-8PR no afecta en nada la radicación oportuna de una planilla de contribución. Por favor, no demore la radicación de su planilla de contribución a la expectativa de una respuesta a su solicitud en la Forma SS-8PR. Usted debe radicar una planilla de contribución para los años en cuestión antes de que se le pueda hacer una determinación relativa a su estado de empleo. Además, si le corresponde, no demore en responder a una solicitud de pago inmediato del saldo debido mientras espera tal determinación.

Instrucciones para Empresas/Pagadores

Si un *trabajador* ha solicitado una determinación de su estado de trabajador mientras trabajaba para usted, usted recibirá una petición del *IRS* para que complete una Forma SS-8PR. En tales casos como éste, solemos otorgar a cada parte en cuestión la oportunidad para presentar una declaración de los hechos ya que nuestra decisión afectará el estado de trabajo para fines contributivos de cada una de las partes en cuestión. Llene esta planilla cuidadosamente. El no responder a esta solicitud no le impedirá que el *IRS* le emita una carta informativa al trabajador, basada en toda la información que éste haya hecho disponible a fin de satisfacer sus obligaciones contributivas federales. Sin embargo, la información que usted provea es de un valor inestimable en la determinación de su estado de trabajador.

Si *usted* está llenando la forma para una clase particular de trabajador, complétela para **un** individuo que es representativo de la clase de trabajadores cuyo estado está en cuestión. Si usted desea una determinación escrita para más de una clase de trabajadores, llene una Forma SS-8PR por separado para un trabajador de cada clase cuyo estado es típico de esa clase. Una determinación por escrito sobre el estado de cualquier trabajador le corresponderá a cada trabajador de la misma clase si los hechos no se difieren substancialmente de los del trabajador cuyo estado de empleo recibió una determinación. Por favor, incluya una lista de los nombres y direcciones de todos los trabajadores que pudieran ser afectados por nuestra determinación.

Si tiene una razón justificante por no tratar a un trabajador suyo como empleado, usted puede ser exonerado de la obligación de pagar las contribuciones por razón del empleo sobre las remuneraciones de ese trabajador según se estipula en la sección 530 de la Ley Contributiva de 1978 (*1978 Revenue Act*). No obstante esto, dichas estipulaciones no se aplican a una determinación solicitada en la Forma SS-8PR ya que la misma no es una revisión o inspección (auditoría) de su planilla de contribución. Para más información sobre la sección 530 de la Ley Contributiva de 1978 y para ver si usted califica para exoneración bajo dicha sección, puede visitar nuestro sitio de la red al *www.irs.gov*.

Aviso sobre la Ley de Confidencialidad de Información y la Ley de Reducción de Trámites

Solicitamos la información contenida en esta forma para cumplir con las leyes que regulan la recaudación de los impuestos internos de los Estados Unidos. Se usará esta información para determinar el estado de empleo del (los) trabajador(es) descrito(s) en esta planilla. El Subtítulo C, las Contribuciones por Razón del Empleo, del Código Federal de Rentas Internas, impone tales contribuciones sobre los sueldos y salarios del empleado. Las secciones 3121(d), 3306(a) y 3401(c) y (d) y su reglamentación respectiva definen "empleado" y "patrono" para propósitos de las contribuciones por razón del empleo impuestas de acuerdo con el Subtítulo C. La sección 6001 le autoriza al *IRS* para solicitar información a fin de determinar si un(os) trabajador(es) o una empresa está(n) o no está(n) sujeta(s) a tales contribuciones. La sección 6109 del Código le requiere que nos provea su número de seguro social o su número de identificación del contribuyente personal. No se le obliga ni al trabajador ni a la empresa que solicite una determinación de su estado de empleo, pero si usted elige hacerlo, tiene que proveer la información solicitada en esta planilla. Si no la provee, a lo mejor, no llegaremos a una determinación apropiada. Si un trabajador o empresa solicita una determinación de su estado de empleo y le pide a usted que provea cierta información para ayudar en la determinación de tal estado, usted no está obligado a someterla. Sin embargo, al no someterla, usted no permitirá al *IRS* llegar a una determinación del estado correcta de ese empleado o empresa. Además, si provee información falsa o fraudulenta, usted puede estar sujeto a pagar multas y/o penalidades. La información facilitada en esta planilla puede ser compartida con el Departamento de Justicia para casos de litigio civil y criminal, y con las ciudades, estados, posesiones y estados asociados libres de los EE.UU. y el Distrito de Columbia para ayudarlos en administrar sus leyes contributivas respectivas. Podemos, además, proporcionarle(s) al (los) trabajador(es) o a la(s) empresa(s) en cuestión esta información como parte de los precedimientos para determinar su estado de empleo.

Usted no está obligado a facilitar la información solicitada en una forma sujeta a la Ley de Reducción de Trámites a menos que la misma muestre un número de control válido de la *OMB*. Los libros o récords relativos a esta forma o sus instrucciones deberán ser conservados mientras su contenido pueda ser utilizado en la administración de cualquier ley contributiva federal. Por regla general, las planillas de contribución y cualquier información pertinente son confidenciales, como lo requiere la sección 6103.

El tiempo que se necesita para llenar y radicar esta forma variará, dependiendo de las circunstancias individuales. El promedio de tiempo que se estima para completar esta forma es: **Mantener los récords,** 22 hr.; **Aprendiendo acerca de la ley o de la forma,** 47 min.; y **Preparar y enviar la forma al IRS,** 1 hr., 11 min.; Si usted desea hacer cualquier comentario acerca de la exactitud de estos estimados o cualquier sugerencia para hacer que esta forma sea más sencilla, por favor, envíenos los mismos. Usted puede escribir al *Tax Forms Committee, Western Area Distribution Center,* Rancho Cordova, CA 95743-0001. **No envíe** esta planilla de contribución a esta dirección. En vez de eso, vea **Adónde se Radica,** en la página 4.

Impreso en papel reciclado

Tests for Exemption Under the Fair Labor Standards Act

Position: _____ Weekly rate of pay: _____

Based on the analysis below, this position is ☐ EXEMPT ☐ NON-EXEMPT

Check each item that is true for that position

For exemption as an EXECUTIVE, all six criteria must be met by the employee in this position:

☐ Has the primary duty of managing the enterprise or a recognized department of it.

☐ Regularly directs the work of at least two or more employees.

☐ Has the authority to hire or fire, or makes recommendations of same that are given great weight.

☐ Customarily and regularly uses discretion in the work done.

☐ Spends at least 80 percent of his or her time on duties directly and closely to the firm's management.

☐ Is paid on a salary basis a minimum of at least $155 per week (in any American state).

Note: If the person is paid at least $250 per week, the employee need only direct two or more employees and perform management duties to be exempt.

For an ADMINISTRATIVE exemption, all six criteria must be met by the employee in this position:

☐ Has the primary duty of office, non-manual work directly related to management policies or general business operations.

☐ Regularly uses discretion and independent judgment and has the authority to make important decisions.

☐ Either regularly assists an executive or performs under general supervision only.

☐ Spends at least 80 percent of his or her time on duties related directly and closely to administration.

☐ Is paid on a salary basis a minimum of at least $155 per week (in any American state).

☐ Performs specialized or technical work requiring special experience, training, or knowledge.

Note: The 80 percent rule does not have to be met if the person is paid at least $250 per week.

For exemption as a PROFESSIONAL, all five criteria must be met by the employee in this position:

☐ Has the primary duty where work requires advanced, specialized learning, or is original and creative resulting from invention or imagination.

☐ Consistently exercises discretion and judgment.

☐ Does work that is primarily intellectual and varied (vs. routine or mechanical).

☐ Spends at least 80 percent of time working on duties related to professional tasks.

☐ Is paid on a salary or fee basis at least $170 per week (in any American state).

Note: The 80 percent rule does not apply to positions paid $250 per week or more.

Analysis Done by: _____ Date: _____

Sample Payroll Deduction Authorization Form

[YOUR COMPANY]

Date:_____

Employee:_____

This memo acknowledges that the employee named above requests [YOUR COMPANY] to deduct $_____ from his or her pay. This amount was created by the following event:

☐ Payroll advance   ☐ Uniforms
☐ Loan to the employee   ☐ Uniform maintenance
☐ Other:_____

This event occurred on the following date:_____

The employee authorizes [YOUR COMPANY] to deduct and retain the following amount from the employee's payroll earnings each pay period until that debt is paid off. The amount is_____.

If any or all of the indebtedness remains unsatisfied upon the employee's termination (for whatever reason) from [YOUR COMPANY], the employee agrees that [YOUR COMPANY] can deduct whatever amount is available and necessary to collect the remaining debt. [YOUR COMPANY] retains the right to pursue any other uncollected debt balance.

By the signature below, the employee voluntarily agrees to enter into this understanding and abide by its terms and conditions.

_____   _____
Employee Signature   Date   Employer Signature   Date

_____
Employee Social Security Number

Notice of Group Health Continuation Coverage Under COBRA

On April 7, 1986, a federal law was enacted (Public Law 99-272, Title X) requiring that most employers sponsoring group health plans offer employees and their families the opportunity for a temporary extension of health coverage (called "continuation coverage") at group rates in certain instances where coverage under the plan would otherwise end. This notice is intended to inform you, in a summary fashion, of your rights and obligations under the continuation coverage pro visions of the law. Both you and your spouse should take the time to read this notice carefully.

If you are an employee of [YOUR COMPANY] covered by the company health plan, you have a right to choose this continuation coverage if you lose your group health coverage because of a reduction in your hours of employment or the termination of your employment (for reasons other than gross misconduct on your part).

If you are the spouse of an employee covered by the company health plan, you have the right to choose continuation coverage for yourself if you lose group health coverage under [Group Health Plan Name] for any of the following four reasons:

- The death of your spouse.
- A termination of your spouse's employment (for reasons other than gross misconduct) or reduction in your spouse's hours of employment with [YOUR COMPANY].
- Divorce or legal separation from your spouse.
- Your spouse becomes entitled to Medicare.

In the case of a dependent child of an employee covered by the company health plan, he or she has the right to continuation coverage if group health coverage under the company health plan is lost for any of the following five reasons:

- The death of the employee.
- A termination of the employee's employment (for reasons other than gross misconduct) or reduction in the employee's hours of employment with the company health plan.
- The employee's divorce or legal separation.
- The employee becomes entitled to Medicare.
- The dependent child ceases to be a "dependent child" under the company health plan.

Under the law, the employee or a family member has the responsibility to inform the company health Plan Administrator of a divorce, legal separation, or a child losing dependent status under the company health plan within 60 days of the date of the later of the event or the date on which coverage would end under the plan because of event. [YOUR COMPANY] has the responsibility to notify the Plan Administrator of the employee's death, termination, reduction in hours of employment or Medicare entitlement. Similar rights may apply to certain retirees, spouses, and dependent children if your employer commences a bankruptcy proceeding and these individuals lose coverage.

[YOUR COMPANY]

Notice of Right to Elect COBRA Continuation and/or Conversion of Group Medical Coverage

(Termination or Reduction in Hours of Employment)

Date of Notice:_____

To: _____
     Employee, Spouse & Dependent Children (if applicable)

_____
Address

_____
City   State   Zip

The _____ group insurance coverage which provides medical care benefits to you (your spouse and dependent children) was or will be discontinued because of the following:

☐ Termination of employment   ☐ Reduction in hours of employment

Under provisions of the Consolidated Omnibus Budget Reconciliation Act of 1985, this is a QUALIFYING EVENT that will entitle you and, if covered by the plan, your spouse and dependent child(ren) ("Qualified Beneficiaries"), if any, to elect to continue coverage (referred to as "COBRA coverage") under the plan for up to 18 months from the date of the qualifying event, **AT YOUR EXPENSE**.

How to Elect COBRA Coverage

Because COBRA gives you the right to elect coverage independently, you, your spouse or dependent children, if any, (hereafter referred to as qualified beneficiaries), will be able to elect single coverage and not include those individuals who do not wish to continue coverage. If you elect to continue coverage you must complete, sign and date the attached election form and send it to the Plan Administrator at the address shown on the last page of the election form by: _____ (60 days from the later of: the date coverage will cease or the date of this notice). Otherwise, your coverage under this plan will cease on _____ (date of qualifying event).

COBRA intends that this notification be given to each of the covered individuals or qualified beneficiaries listed above. However, the law allows for one notice addressed to all covered individuals to be sent to a single address. Should any of the eligible individuals listed above reside at a different address, you must advise us immediately at the address below so that we may send them a notice of election form. However, only one of you needs to elect continuation coverage for your spouse and dependent child(ren) who wish to continue coverage. Please note that your spouse and dependent children are not eligible to elect continuation coverage unless they were covered by this plan at the time of the qualifying event. However, COBRA continuation coverage will also be available for dependent children who are born to or placed for adoption with a covered employee after termination of coverage under the plan if the covered employee elected to continue coverage.

Payment of COBRA Coverage Premiums

If you elect to continue coverage, you must, within 45 days of your election date, submit to the plan administrator your check to cover the initial premium payment. The check for this initial premium payment must cover the number of full months from the date of termination, _____, to the time of your payment. After the initial payment, you must submit the monthly payment on the first of each month. Your monthly premium amount may change in the future. You will be advised of a change in the premium amount that applies to all participants in this plan and the date you are required to pay the new premium amount. The current amount of the premium for your coverage is explained on the attached form.

Length of COBRA Coverage Period

If a qualified beneficiary elects coverage, it will last for as long as **18 months** beginning on the date of the covered employee's qualifying event. This period may be extended for the following reasons:

Death of employee, divorce, legal separation or change in dependent status. If these events occur during the original 18-month period of coverage, the period of coverage for a qualified beneficiary may be extended for an additional 18 months, resulting in a total of 36 months of coverage from the date of the original qualifying event. Note that to receive this extension, you and/or spouse and dependent child(ren) must notify the plan administrator within 60 days of the occurrence of these events.

Medicare entitlement of employee. If you become entitled to Medicare before the your qualifying event, your spouse or dependent child(ren) if any may receive extended COBRA coverage for up to the greater of either: (a) 36 months from the date of the covered employee's Medicare entitlement; or (b) 18 months from the date of the covered employee's qualifying event.

If you become entitled to Medicare after your qualifying event but within 18 months of your qualifying event, your spouse and dependent child(ren), if any, may receive an additional 18 months of COBRA coverage. Note that a person generally has become entitled to Medicare when he or she has applied for Social Security income payments or has filed an application for benefits under Part A or Part B of Medicare.

Disability Extension. If a qualified beneficiary is determined to be disabled by the Social Security Administration at any time during the first 60 days of COBRA coverage, the disabled individual may extend the 18-month period for an additional 11 months, resulting in a total of 29 months from the date of the original qualifying event. This extension only applies if the Plan Administrator is notified within 60 days of a disability determination and before the end of the 18-month period. It is the responsibility of the affected individual to obtain the disability determination from the Social Security Administration and to provide a copy of the determination to the plan administrator within this notification period. If there is a final determination that the qualified beneficiary is no longer disabled, you must notify the plan administrator of such within 30 days of the determination.

Bankruptcy filing. If [YOUR COMPANY] files for bankruptcy reorganization and retiree health coverage is lost within one year before or after the bankruptcy filing, COBRA coverage could continue until the death of a retiree (or a surviving spouse of a deceased retiree) or for 36 months from the retiree's death (after the bankruptcy filing) in the case of the spouse and dependent child(ren).

Early Termination of COBRA Coverage

COBRA coverage may terminate early if:

- The required premium payment is not paid when due.
- A qualified beneficiary becomes covered under another group health plan that does not contain any exclusion or limitation for any that individual's pre-existing conditions.
- A qualified beneficiary becomes entitled to Medicare benefits.
- All of the company's group health plans are terminated.

- Coverage is extended to 29 months due to disability, and a determination is made that the individual is no longer disabled. *Note:* Federal law requires that you inform the Plan Administrator of any final determination that the individual is no longer disabled within 30 days of such determination.
- A qualified beneficiary notifies the plan administrator that he/she wishes to cancel continuation coverage.

IMPORTANT

Continuation coverage under COBRA is provided subject to the eligibility of each of the qualified beneficiaries. The Plan Administrator reserves the right to terminate your COBRA coverage retroactively if a qualified beneficiary is determined to be ineligible for coverage. To be sure that all qualified beneficiaries receive the necessary information concerning your rights, you should keep the Plan Administrator informed of any address changes. This notice is a summary of your COBRA rights. For answers to specific questions, please contact the Plan Administrator at the telephone number or address listed below.

Name of Plan Administrator: _____

Address: _____

Phone Number: _____

Notice of Your Right to Documentation of Health Coverage

Recent changes in federal law may affect your health coverage if you are enrolled or become eligible to enroll in a health plan that excludes coverage for pre-existing conditions. The Health Insurance Portability and Accountability Act of 1996 (HIPAA) limits the circumstances under which coverage may be excluded in a subsequent health plan for medical conditions present before you enroll. Under the law, a pre-existing condition exclusion generally may not be imposed for more than 12 months (18 months for late enrollment). The 12-month (or 18-month) exclusion period is reduced by your prior health coverage. You are entitled to a certificate that will show evidence of your prior health coverage. If you buy health insurance other than through an employer group health plan, a certificate of prior coverage may help you obtain coverage without a pre-existing condition exclusion. Contact your state insurance department for further information.

For employer group health plans, these changes generally take effect at the beginning of the first plan year starting after June 30, 1997. For example, if your employer's plan year begins on January 1, 1998, the plan is not required to give you credit for your prior coverage until January 1, 1998.

You have the right to receive a certificate of prior health coverage since July 1, 1996. You may need to provide other documentation for earlier period of health care coverage. Check with your new plan administrator to see if your new plan excludes coverage for preexisting conditions and if you need to provide a certificate or other documentation of your previous coverage.

To get a certificate, complete the attached form and return it to:

[COMPANY]

[Address]

For additional information, contact: [Name and Telephone Number]

* * * * * *

Request for Certificate of Health Coverage

Name of Participant: _____ Date: _____

Address: _____

Phone Number: _____

Name and relationship of any dependents for whom certificates are requested (please provide addresses if different from above):

_____

_____

_____

_____

[YOUR COMPANY]

COBRA Continuation Coverage Election Form
(Termination or Reduction in Hours of Employment)

Date of Notice: _____

☐ Mailed
☐ Hand Delivered

Qualified Beneficiary Information

Name: Last, First, Middle   Social Security Number

_____

Home Address   Street   City   State   Zip

_____

Date of Birth: _____ /_____ /_____   Marital Status: ☐ Single ☐ Married

No. of Dependent Children: _____

Date of Hire: _____ /_____ /_____   Policy Number: _____

Entitlement to COBRA Coverage

As explained in the notice of rights accompanying this form, you and, if covered by the plan, your spouse and dependent child(ren), if applicable, could be entitled to continue health coverage under the company's group health plan due to the following qualifying event, which is effective

_____:

☐ Termination of employment   ☐ Reduction in hours of employment

This qualifying event will result in the loss of health coverage for all qualified beneficiaries unless you elect continuation coverage. If you would like to elect continuation coverage, please read and sign this form and return it to the address below as soon as possible.

If this election form is not returned within 60 days of the later of the date coverage will cease or the date of this notice, you will lose your right to elect coverage, and your coverage under the company's group health plan will terminate effective _____. Enrollment coverage will be the same for all family members unless a separate COBRA Election Form is furnished to the Plan Administrator for each family member.

Continuation coverage under COBRA is provided subject to your eligibility. The Plan Administrator reserves the right to terminate your COBRA coverage retroactively if you are determined to be ineligible for coverage.

If you do not return this election form within 60 days of the later of the date coverage will cease or the date of this letter you will lose your right to elect continuation coverage.

Length of COBRA Coverage

You and any qualified beneficiary (spouse and dependent child(ren)) are eligible to receive up to **18 months** of continuation coverage from the date of termination or reduction of hours of employment. However, coverage may extend beyond that period or terminate early, as explained in your election notice.

COBRA Coverage Premiums

Within 45 days after the date that you elect COBRA coverage, you must pay an initial premium, which includes:

- The period of coverage from the date of your qualifying event to the date of your election.
- Any regularly scheduled monthly premium that becomes due between your election and the end of the 45-day period.

Once the Plan Administrator receives this election form, you will be notified of the amount of the initial premium you must pay. **If you fail to pay the initial premium, or any subsequent monthly premium, in a timely fashion, your coverage will terminate. Checks are to be made payable to _____.**

Premium payments are generally due within 30 days after the first day of each month of coverage. Premium amounts change from time to time. You will be notified of any change in the premium amount.

The current monthly cost for COBRA continuation coverage under the plan for the different levels of coverage is as follows:

Individual $ _____

Husband & Wife $ _____

Parent & Child $ _____

Family $ _____

You currently have_____ coverage under the Company's plan. You are eligible to continue the coverage you had immediately prior to the occurrence of the qualifying event and unless you expressly elect otherwise, that coverage will be continued for you and any of the qualified beneficiaries covered by the plan as well. **If premium payment is not received on time, coverage will terminate and may not be reinstated.**

COBRA Coverage Election Agreement

I have read this form and the notice of my election rights. I understand my rights to elect continuation coverage and would like to take the action indicated below. I understand that if I elect continuation coverage and I fail to pay any premium payment on time, this coverage will terminate. I also agree to notify the Plan Administrator if I or any member of my family become(s) covered under another group health plan or entitled to Medicare.

Please check ONE only.

☐ **I elect to continue individual coverage under the plan.**

Name	Relationship	Birth Date	Soc. Sec. #

☐ **I elect to continue husband/wife coverage under the plan.**

Name	Relationship	Birth Date	Soc. Sec. #

☐ **I elect to continue parent/child coverage under the plan.**

Name	Relationship	Birth Date	Soc. Sec. #

☐ **I elect to continue family coverage under the plan. (Only to be checked by those qualified beneficiaries who had family coverage before the qualifying event, or in the event that a dependent child is born to or placed for adoption with a covered employee).**

List dependents to be covered:

Name	Relationship	Birth Date	Soc. Sec. #

☐ **I have read this form and the notice of rights. I am waiving my right to continuation coverage under the plan. (One form for each qualified beneficiary waiving his/her right to continuation coverage must be completed and forwarded to the plan administrator).**

Signature: _____ Date: _____

Name (Please Print): _____

Address: _____

Telephone: _____

Received By Plan Administrator: _____

Date: _____

If you elect to continue coverage, please return this signed election form to:

Name of Administrator: _____

Address: _____

Phone Number: _____

Your first monthly payment for continuation coverage must be enclosed with this notice or submitted within 45 days of your election pursuant to this notice. Subsequent monthly payments should be received by us the 1st day of each month. There is a grace period of at least 30 days for payment of a regularly scheduled premium.

IMPORTANT

The group insurance Plan under which coverage for you and any of the qualified beneficiaries being continued may contain a conversion privilege that permits applying for an individual policy of medical care benefits. If you elect to continue coverage under the group Plan, application for such an individual policy may also be made upon cessation of the continued coverage, during or at the end of the applicable continuation period. If you do not elect to continue coverage under the group insurance plan, you may also apply for an individual policy. If you are interested in making such a conversion, please contact your group Plan Administrator for more details.

[YOUR COMPANY]

Notice of Right to Elect COBRA Continuation Coverage
(Death, Divorce, Legal Separation, Medicare Entitlement)

Date of Notice:_____

To: _____
  Name of Spouse and/or Dependents

_____
Address

_____
City   State   Zip

Name of Employee: _____

On _____ , the Plan Administrator of _____
group health plan was notified that your health coverage (and that of your dependent child(ren), if any)
will terminate because of the following:

☐ Your spouse's death
☐ Your divorce or legal separation
☐ Your spouse's entitlement to Medicare

Under provisions of the Consolidated Omnibus Budget Reconciliation Act of 1985 ("COBRA"), this is a
QUALIFYING EVENT that will entitle you and your dependent child(ren), if any, to elect to continue
coverage (referred to as "COBRA coverage") under the Plan for up to 36 months from the date of your
qualifying event, **AT YOUR EXPENSE**.

How to Elect COBRA Coverage

To continue coverage, you must complete, sign, and date the attached election form and send it to the
Plan Administrator at the address shown on the last page of this form by _____
sixty (60) days from the later of the date coverage will cease or the date of this notice). Otherwise, your
coverage under this plan will cease on _____ (date of qualifying event).

COBRA intends that this notification be given to each of the covered individuals or qualified beneficiaries
listed above. However, the law allows for one notice addressed to all covered individuals to be sent to a
single address. Should any of the eligible individuals listed above reside at a different address, you must
advise us immediately at the address below so that we may send them a notice of election form.
However, only one of you needs to elect continuation coverage for all the qualified beneficiaries who
wish to continue coverage. Unless you state otherwise, your election will include any dependent
child(ren) who will lose coverage because of the qualifying event. However, because COBRA gives you
the right to elect coverage independently, you or your dependent child(ren) may elect single coverage
and not include those individuals who do not wish to continue coverage. You and your dependent
children are not eligible to elect continuation coverage unless covered by the plan at the time of the
qualifying event. However, COBRA continuation coverage will also be available for dependent children
who are born to or placed for adoption with a covered employee after termination of coverage under
the plan if the covered employee elected to continue coverage.

Payment of COBRA Premiums

If you elect to continue coverage, you must, within 45 days of your election date, submit to the plan administrator your check to cover the initial premium payment. The check for this initial premium payment must cover the number of full months from the date of termination, _____, to the time of your payment. After the initial payment, you must submit the monthly payment on the first of each month. You monthly premium amount may change in the future. You will be advised of a change in the premium amount that applies to all participants in this plan and the date you are required to pay the new premium amount. The current amount of the premium for your coverage is explained on the attached form.

Length of COBRA Coverage Period

If you elect coverage, it will last for as long as **36 months** beginning on the date of your qualifying event. However, if the employer files for bankruptcy reorganization and retiree health coverage is lost within one year before or after the bankruptcy filing, COBRA coverage could continue until the death of a retiree (or the surviving spouse of a deceased retiree) or for 36 months from the retiree's death (after the bankruptcy filing) in the case of the spouse and dependent child(ren).

Early Termination of COBRA Coverage

The COBRA coverage of you and your dependent child(ren) may be terminated early if:

- The required premium payment is not paid when due.
- You or your dependent child(ren) become covered under another group health plan that does not contain any exclusion or limitation for any of your pre-existing conditions.
- You or your dependent child(ren) become entitled to Medicare benefits. A person generally has become entitled to Medicare when he or she has applied for Social Security income payments or has filed an application for benefits under Part A or Part B of Medicare.
- All of the company's group health plans are terminated.
- A qualified beneficiary notifies the plan administrator that he/she wishes to cancel continuation coverage.

IMPORTANT

Continuation coverage under COBRA is provided subject to the eligibility of each of the qualified beneficiaries. The Plan Administrator reserves the right to terminate your COBRA coverage retroactively if a qualified beneficiary is determined to be ineligible for coverage. To be sure that you and your dependent children, if any, receive the necessary information concerning your rights, you should keep the Plan Administrator informed of any address changes. This notice is a summary of your COBRA rights. If you have any specific questions, please contact the Plan Administrator at the telephone number or address listed below.

Name of Plan Administrator: _____

Address: _____

Phone Number: _____

Notice of Your Right to Documentation of Health Coverage

Recent changes in federal law may affect your health coverage if you are enrolled or become eligible to enroll in a health plan that excludes coverage for pre-existing conditions. The Health Insurance Portability and Accountability Act of 1996 (HIPAA) limits the circumstances under which coverage may be excluded in a subsequent health plan for medical conditions present before you enroll. Under the law, a pre-existing condition exclusion generally may not be imposed for more than 12 months (18 months for late enrollment). The 12-month (or 18-month) exclusion period is reduced by your prior health coverage. You are entitled to a certificate that will show evidence of your prior health coverage. If you buy health insurance other than through an employer group health plan, a certificate of prior coverage may help you obtain coverage without a pre-existing condition exclusion. Contact your state insurance department for further information.

For employer group health plans, these changes generally take effect at the beginning of the first plan year starting after June 30, 1997. For example, if your employer's plan year begins on January 1, 1998, the plan is not required to give you credit for your prior coverage until January 1, 1998.

You have the right to receive a certificate of prior health coverage since July 1, 1996. You may need to provide other documentation for earlier period of health care coverage. Check with your new plan administrator to see if your new plan excludes coverage for pre-existing conditions and if you need to provide a certificate or other documentation of your previous coverage.

To get a certificate, complete the attached form and return it to:

[COMPANY]

[Address]

For additional information, contact: [Name and Telephone Number]

* * * * * *

Request for Certificate of Health Coverage

Name of Administrator: _____

Address: _____

Phone Number: _____

Name and relationship of any dependents for whom certificates are requested (please provide addresses if different from above):

COBRA Continuation Coverage Election Form

(Death, Divorce, Legal Separation, Medicare Entitlement)

Date of Notice: _____

☐ Mailed
☐ Hand Delivered

Qualified Beneficiary Information

Name: Last, First, Middle   Social Security Number

_____

Home Address   Street   City   State   Zip

_____

Date of Birth: _____ / _____ / _____   Marital Status: ☐ Single ☐ Married

No. of Dependent Children: _____

Date of Hire: _____ / _____ / _____   Policy Number: _____

Entitlement to COBRA Coverage

As explained in the notice of rights accompanying this form, you are entitled to continue health coverage under the company's group health plan due to the following qualifying event, which became effective _____:

☐ Your spouse's death
☐ Your divorce or legal separation
☐ Your spouse's entitlement to Medicare

This qualifying event will result in the loss of your health care coverage unless you elect continuation coverage. If you would like to elect continuation coverage, please read and sign this form and return it to the address below as soon as possible.

If you do not return this election form within 60 days of the later of the date coverage will cease or the date of this notice, you will lose your right to elect coverage, and your coverage under the company's group health plan will terminate effective _____.

Continuation coverage under COBRA is provided subject to your eligibility. The Plan Administrator reserves the right to terminate your COBRA coverage retroactively if you are determined to be ineligible for coverage.

If you do not return this election form within 60 days from the later of the date coverage will cease or the date of this notice, you will lose your right to elect continuation coverage.

Length of COBRA Coverage

You are eligible to receive up to **36 months** of continuation coverage from the date of the qualifying event. However, coverage may terminate early, as explained in your election form.

COBRA Coverage Premiums

Within 45 days after the date that you elect COBRA coverage, you must pay an initial premium, which includes:

- The period of coverage from the date of your qualifying event to the date of your election.
- Any regularly scheduled monthly premium that becomes due between your election and the end of the 45-day period.

Once the Plan Administrator receives this election form, you will be notified of the amount of the initial premium you must pay. **If you fail to pay the initial premium, or any subsequent monthly premium, in a timely fashion, your coverage will terminate. Checks are to be made payable to _____.**

You currently have _____ coverage under the Company's plan. You may elect individual, parent/child or family coverage. Unless you expressly elect otherwise, your current level of coverage will be continued for you and your children, if any. The current monthly cost for COBRA continuation coverage under the plan is a follows:

Individual $ _____

Parent & Child $ _____

Family $ _____

Premium payments are generally due within 30 days after the first day of each month of coverage. Premium amounts change from time to time. You will be notified of any change in this premium amount. **If premium payment is not received on time, coverage will terminate and may not be reinstated.**

COBRA Coverage Election Agreement

I have read this form and the notice of my election rights. I understand my rights to elect continuation coverage and would like to take the action indicated below. I understand that if I elect continuation coverage and I fail to pay any premium payment on time, this coverage will terminate. I also agree to notify the Plan Administrator if I (or any member of my family) become(s) covered under another group health plan or entitled to Medicare.

Please check ONE option only.

☐ **I elect to continue individual coverage under the plan.**

    Name   Relationship   Birth Date   Soc. Sec. #

_____

☐ **I elect to continue parent/child coverage under the plan.**

    Name   Relationship   Birth Date   Soc. Sec. #

_____

_____

☐ **I elect to continue family coverage under the plan. (Only to be checked by those qualified beneficiaries who had family coverage before the qualifying event, or in the event that a dependent child is born to or placed for adoption with a covered employee).**

List dependents to be covered:

    Name   Relationship   Birth Date   Soc. Sec. #

_____

_____

_____

_____

☐ **I have read this form and the notice of rights. I am waiving my right to continuation coverage under the plan. (One form for each qualified beneficiary waiving his/her right to continuation coverage must be completed and forwarded to the plan administrator).**

Signature:_____ Date:_____

Name (Please Print):_____

Address:_____

Telephone:_____

Received By Plan Administrator:_____

Date:_____

If you elect to continue coverage, please return this signed election form to:

Name of Administrator:_____

Address:_____

Phone Number:_____

Your first monthly payment for continuation coverage must be enclosed with this notice or submitted within 45 days of your election pursuant to this notice. Subsequent monthly payments should be received by us the 1st day of each month. There is a grace period of at least 30 days for payment of a regularly scheduled premium.

IMPORTANT

The group insurance Plan under which coverage for you and, if applicable, your covered dependents, if any, is being continued also contains a conversion privilege which permits applying for an individual policy of medical care benefits. If you elect to continue coverage under the group Plan, application for such an individual policy may also be made upon cessation of the continued coverage, during or at the end of the applicable continuation period. If you do not elect to continue coverage under the group insurance plan you may also apply for an individual policy. If you are interested in making such a conversion, please contact your group Plan Administrator for more details.

[YOUR COMPANY]

Notice of Right to Elect COBRA Continuation Coverage
(Cessation of Dependent Status of Dependent Child)

Date of Notice: _____

To: _____
Name of Spouse and/or Dependents

_____
Address

_____
City   State   Zip

On_____ , the Plan Administrator of _____ group
health plan was notified that your health coverage under the plan will terminate because you have
ceased to be a dependent child as defined under the plan:

Under provisions of the Consolidated Omnibus Budget Reconciliation Act of 1985, this is a QUALIFYING
EVENT that will entitle you to elect to continue coverage (referred to as "COBRA coverage") under the
plan for up to 36 months from the date of your qualifying event, **AT YOUR EXPENSE**.

How to Elect COBRA Coverage

To continue coverage, you or your parent(s) must complete and submit the attached election form to
the Plan Administrator by: _____ (sixty (60) days from the later of: the date
coverage will cease or the date of this notice). Otherwise, your coverage under this plan will cease on
_____ (date of qualifying event).

Payment of COBRA Premiums

If you elect to continue coverage, you must, within 45 days of your election date, submit to the plan
administrator your check to cover the initial premium payment. The check for this initial premium payment must cover the number of full months from the date of termination, _____,
to the time of your payment. After the initial payment, you must submit the monthly payment on the
first of each month. Your monthly premium amount may change in the future. You will be advised of a
change in the premium amount that applies to all participants in this plan and the date you are required
to pay the new premium amount. The current amount of the premium for your coverage is explained
on the attached form.

Length of COBRA Coverage Period

If you elect coverage, it will last for as long as **36 months** beginning on the date that you ceased to be a dependent child under the plan.

Early Termination of COBRA Coverage

Your COBRA coverage may be terminated early for a number of reasons, including:

- You or your parents do not pay the required premium on time.
- You become covered under another group health plan that does not contain any exclusion or limitation for any of your pre-existing conditions.
- You become entitled to Medicare benefits. A person generally has become entitled to Medicare when he or she has applied for Social Security income payments or has filed an application for benefits under Part A or Part B of Medicare.
- All of the company's group health plans are terminated.
- You or your parent(s) notify the plan administrator that you wish to cancel continuation coverage.

IMPORTANT

Continuation coverage under COBRA is provided subject to the eligibility of each of the qualified beneficiaries. The Plan Administrator reserves the right to terminate your COBRA coverage retroactively if a qualified beneficiary is determined to be ineligible for coverage. To be sure that all qualified beneficiaries receive the necessary information concerning your rights, you should keep the Plan Administrator informed of any address changes. This notice is a summary of your COBRA rights. For answers to specific questions, please contact the Plan Administrator at the telephone number or address listed below.

Name of Plan Administrator: _____

Address: _____

Phone Number: _____

Notice of Your Right to Documentation of Health Coverage

Recent changes in federal law may affect your health coverage if you are enrolled or become eligible to enroll in a health plan that excludes coverage for pre-existing conditions. The Health Insurance Portability and Accountability Act of 1996 (HIPAA) limits the circumstances under which coverage may be excluded in a subsequent health plan for medical conditions present before you enroll. Under the law, a pre-existing condition exclusion generally may not be imposed for more than 12 months (18 months for late enrollment). The 12-month (or 18-month) exclusion period is reduced by your prior health coverage. You are entitled to a certificate that will show evidence of your prior health coverage. If you buy health insurance other than through an employer group health plan, a certificate of prior coverage may help you obtain coverage without a preexisting condition exclusion. Contact your state insurance department for further information.

For employer group health plans, these changes generally take effect at the beginning of the first plan year starting after June 30, 1997. For example, if your employer's plan year begins on January 1, 1998, the plan is not required to give you credit for your prior coverage until January 1, 1998.

You have the right to receive a certificate of prior health coverage since July 1, 1996. You may need to provide other documentation for earlier period of health care coverage. Check with your new plan administrator to see if your new plan excludes coverage for preexisting conditions and if you need to provide a certificate or other documentation of your previous coverage.

To get a certificate, complete the attached form and return it to:

[COMPANY]

[Address]

For additional information, contact: [Name and Telephone Number]

* * * * * *

Request for Certificate of Health Coverage

Name of Participant: _____ Date: _____

Address: _____

Phone Number: _____

Name and relationship of any dependents for whom certificates are requested (please provide addresses if different from above):

_____

_____

_____

_____

[YOUR COMPANY]

COBRA Continuation Coverage Election Form

(Cessation of Dependent Status of Dependent Child)

Date of Notice: _____

Qualified Beneficiary Information

Name: Last, First, Middle	Social Security Number
Home Address Street	City State Zip
Policy Number: _____	Date of Birth: _____ / _____ / _____
Parent's Names: Employee_____	Spouse_____
Address, if different: _____	

Entitlement to COBRA Coverage

As explained in the notice of rights accompanying this form, you may be entitled to continue health coverage under the company's group health plan because you have ceased to be a dependent as defined by the plan. The effective date of this qualifying event is: _____.

This qualifying event will result in the loss of your health coverage unless you elect continuation coverage. If you would like to elect continuation coverage, please read and sign this form and return it to the address below as soon as possible.

If you do not return this election form within 60 days of the date of this notice (or the date coverage will cease, whichever is later), you will lose your right to elect coverage, and your coverage under the company's group health plan will terminate effective: _____.

Continuation coverage under COBRA is provided subject to your eligibility. The Plan Administrator reserves the right to terminate your COBRA coverage retroactively if you are determined to be ineligible for coverage.

If you do not return this election form within 60 days of the date of this letter or the date your coverage ceases, which ever is later, you will lose your right to elect continuation coverage.

Length of COBRA Coverage

You could be eligible to receive up to **36 months** of continuation coverage from the date of your qualifying event. However, coverage may terminate early, as explained in your election form.

COBRA Coverage Premiums

Within 45 days after the date that you elect COBRA coverage, you must pay an initial premium, which includes:

- The period of coverage from the date of your qualifying event to the date of your election.
- Any regularly scheduled monthly premium that becomes due between your election and the end of the 45-day period.

Once the Plan Administrator receives this election form, you will be notified of the amount of the premium you must pay. **If you fail to pay the initial premium, or any subsequent monthly premium, in a timely fashion, your coverage will terminate. Checks are to be made payable to _____.** You are eligible for individual coverage. The current monthly cost for COBRA continuation coverage is $_____.

Premium amounts change from time to time. You will be notified of any change in this premium amount. Premium payments are generally due within 30 days after the first day of each month of coverage. **If premium payment is not received on time, coverage will terminate and may not be reinstated.**

COBRA Coverage Election Agreement

I have read this form and the notice of my election rights. I understand my rights to elect continuation coverage and would like to take the action indicated below. I understand that if I elect continuation coverage and I fail to pay any premium payment on time, this coverage will terminate. I also agree to notify the Plan Administrator if I become covered under another group health plan or entitled to Medicare.

Please check ONE option only:

☐ **I elect to continue individual coverage under the plan.**

☐ **I have read this form and the notice of rights. I am waiving my right to continuation coverage under the plan.**

Signature _____ Date: _____

Name (Please Print): _____

Address: _____

Telephone: _____

Received By Plan Administrator: _____ Date: _____

If you elect to continue coverage, please return this signed election form to:

Name of Administrator: _____

Address: _____

Phone Number: _____

Your first monthly payment for continuation coverage must be enclosed with this notice or submitted within 45 days of your election pursuant to this notice. Subsequent monthly payments should be received by us the 1st day of each month. There is a grace period of at least 30 days for payment of a regularly scheduled premium.

IMPORTANT

The group insurance Plan under which coverage for you is being continued may contain a conversion privilege which permits applying for an individual policy of medical care benefits. If you elect to continue coverage under the group Plan, application for such an individual policy may also be made upon cessation of the continued coverage, during or at the end of the applicable continuation period. If you do not elect to continue coverage under the group insurance plan, you may also apply for an individual policy. If you are interested in making such a conversion, please contact your group plan administrator for more details.

[YOUR COMPANY]

Notice of Right to Elect COBRA Continuation Coverage
(Retirement)

Date of Notice: _____

To: _____
Employee, Spouse & Dependent Children (if applicable)

_____
Address

_____
City   State   Zip

On _____, the Plan Administrator of the company's group health plan was notified that your health coverage, as well as that of your spouse and dependent child(ren), if any (hereafter collectively referred to as qualified beneficiaries) under the plan will terminate because of your retirement, which is effective _____.

Under provisions of the Consolidated Omnibus Budget Reconciliation Act of 1985, this is a QUALIFYING EVENT that entitles qualified beneficiaries to elect to continue coverage (referred to as "COBRA coverage") under the plan for up to 18 months from the date of your retirement, **AT YOUR EXPENSE**.

Electing COBRA Coverage

Because COBRA gives you the right to elect coverage independently, you, your spouse, or dependent children, if any, (hereafter referred to as qualified beneficiaries), will be able to elect single coverage and not include those individuals who do not wish to continue coverage. If you elect to continue coverage you must complete, sign and date the attached election form and send it to the Plan Administrator at the address shown on the last page of the election form by: _____ (60 days from the later of: the date coverage will cease or the date of this notice). Otherwise, your coverage under this plan will cease on _____ (date of qualifying event).

COBRA intends that this notification be given to each of the covered individuals or qualified beneficiaries listed above. However, the law allows for one notice addressed to all covered individuals to be sent to a single address. Should any of the eligible individuals listed above reside at a different address, you must advise us immediately at the address below so that we may send them a notice of election form. However, only one of you needs to elect continuation coverage for your spouse and dependent child(ren) who wish to continue coverage. Please note that your spouse and dependent children are not eligible to elect continuation coverage unless they were covered by this plan at the time of the qualifying event. However, COBRA continuation coverage will also be available for dependent children who are born to or placed for adoption with a covered employee after termination of coverage under the plan if the covered employee elected to continue coverage.

Payment of COBRA Coverage Premiums

If you elect to continue coverage, you must, within 45 days of your election date, submit to the plan administrator your check to cover the initial premium payment. The check for this initial premium payment must cover the number of full months from the date of termination, _____, to the time of your payment. After the initial payment, you must submit the monthly payment on the first of each month. Your monthly premium amount may change in the future. You will be advised of a change in the premium amount that applies to all participants in this plan and the date you are required to pay the new premium amount. The current amount of the premium for your coverage is explained on the attached form.

Length of COBRA Coverage Period

A qualified beneficiary may elect COBRA coverage, which will last for up to **18 months** from the date of the retirement. This period may be extended for the following reasons:

Death of employee, divorce, legal separation or change in dependent status. If these events occur during the original 18-month period, the period of coverage for a qualified beneficiary may be extended for an additional 18 months, resulting in a total of 36 months of coverage from the date of your retirement. Note that, to receive this extension, you and/or spouse and dependent child(ren) must notify the plan administrator within 60 days of the occurrence of these events.

Medicare entitlement of employee. If you become entitled to Medicare before your retirement, your spouse and dependent child(ren), if any, may receive extended COBRA coverage for up to the greater of either: (a) 36 months from the date of your Medicare entitlement; or (b) 18 months from the date of your retirement.

If you become entitled to Medicare after your retirement but within 18 months of the date of retirement, a qualified beneficiary, may receive an additional 18 months of COBRA coverage. Note that a person generally has become entitled to Medicare when he or she has applied for Social Security income payments or has filed an application for benefits under Part A or Part B of Medicare.

Disability determination. If a qualified beneficiary is determined to be disabled under the Social Security Act at any time during the first 60 days of COBRA coverage, the 18-month period is extended to 29 months from the date of your retirement. This extension only applies if you notify the Plan Administrator within 60 days of a disability determination and before the end of the 18-month period. It is the responsibility of the affected individual to obtain the disability determination from Social Security Administration and to provide a copy of the determination to the plan administrator within this notification period. If there is a final determination that the qualified beneficiary is no longer disabled, you must notify the plan administrator of such within 30 days of the determination.

Bankruptcy filing. If the employer files for bankruptcy reorganization and retiree health coverage is lost within one year before or after the bankruptcy filing, COBRA coverage may continue until the death of a retiree (or a surviving spouse of a deceased retiree) or for 36 months from a retiree's death (after the bankruptcy filing) in the case of the spouse and dependent child(ren).

Early Termination of COBRA Coverage

COBRA coverage may terminate early for a number of reasons, including:

- Your premium is not paid on time.
- A qualified beneficiary becomes covered under another group health plan that does not contain any exclusion or limitation for any of your pre-existing conditions.

- A qualified beneficiary becomes entitled to Medicare benefits.
- All of the company's group health plans are terminated.
- Coverage is extended to 29 months due to disability, and a determination is made that the individual is no longer disabled. *Note:* Federal law requires that you inform the Plan Administrator of any final determination that the individual is no longer disabled within 30 days of such determination.
- A qualified beneficiary notifies the plan administrator that he/she wishes to cancel continuation coverage.

IMPORTANT

Continuation coverage under COBRA is provided subject to the eligibility of each of the qualified beneficiaries. The Plan Administrator reserves the right to terminate your COBRA coverage retroactively if a qualified beneficiary is determined to be ineligible for coverage. To be sure that all qualified beneficiaries receive the necessary information concerning your rights, you should keep the Plan Administrator informed of any address changes. This notice is a summary of your COBRA rights. For answers to specific questions, please contact the Plan Administrator at the telephone number or address listed below.

Name of Plan Administrator: _____

Address: _____

Phone Number: _____

Notice of Your Right to Documentation of Health Coverage

Recent changes in federal law may affect your health coverage if you are enrolled or become eligible to enroll in a health plan that excludes coverage for pre-existing conditions. The Health Insurance Portability and Accountability Act of 1996 (HIPAA) limits the circumstances under which coverage may be excluded in a subsequent health plan for medical conditions present before you enroll. Under the law, a pre-existing condition exclusion generally may not be imposed for more than 12 months (18 months for late enrollment). The 12-month (or 18-month) exclusion period is reduced by your prior health coverage. You are entitled to a certificate that will show evidence of your prior health coverage. If you buy health insurance other than through an employer group health plan, a certificate of prior coverage may help you obtain coverage without a pre-existing condition exclusion. Contact your State insurance department for further information.

For employer group health plans, these changes generally take effect at the beginning of the first plan year starting after June 30, 1997. For example, if your employer's plan year begins on January 1, 1998, the plan is not required to give you credit for your prior coverage until January 1, 1998.

You have the right to receive a certificate of prior health coverage since July 1, 1996. You may need to provide other documentation for earlier period of health care coverage. Check with your new plan administrator to see if your new plan excludes coverage for pre-existing conditions and if you need to provide a certificate or other documentation of your previous coverage.

To get a certificate, complete the attached form and return it to:

[COMPANY]

[Address]

For additional information contact: [Name and Telephone Number]

* * * * * *

Request for Certificate of Health Coverage

Name of Participant: _____ Date: _____

Address: _____

Phone Number: _____

Name and relationship of any dependents for whom certificates are requested (please provide addresses if different from above):

_____

_____

_____

_____

[YOUR COMPANY]

COBRA Continuation Coverage Election Form
(Retirement)

Date of Notice: _____

☐ Mailed
☐ Hand Delivered

Qualified Beneficiary Information

Name: Last, First, Middle	Social Security Number

Home Address Street	City	State	Zip

Date of Birth: _____ /_____ /_____   Marital Status: ☐ Single ☐ Married

No. of Dependent Children: _____

Date of Hire: _____ /_____ /_____   Policy Number: _____

Entitlement to COBRA Coverage

As explained in the notice of rights accompanying this form, you and your spouse and dependent child(ren), if any, could be entitled to elect COBRA coverage under the company's group health plan due to your retirement, which is effective _____.

This qualifying event will result in the loss of your health coverage unless you elect continuation coverage. If you would like to elect continuation coverage, please read and sign this form and return it to the address below as soon as possible.

If you do not return this election form within 60 days from the later of the date coverage will cease or the date of this notice, you will lose your right to elect coverage, and your coverage under the company's group health plan will terminate effective: _____.

Continuation coverage under COBRA is provided subject to your eligibility. The Plan Administrator reserves the right to terminate your COBRA coverage retroactively if you are determined to be ineligible for coverage.

If you do not return this election form within 60 days of the date of this letter or the date your coverage ceases, whichever is later, you will lose your right to elect continuation coverage.

Length of COBRA Coverage

As explained in the notice accompanying this form, you and your spouse and/or dependent child(ren), are entitled to **18 months** of COBRA coverage. However, COBRA coverage may extend beyond the 18-month period or terminate early.

COBRA Coverage Premiums

Within 45 days after the date that you elect COBRA coverage, you must pay an initial premium, which includes:

- The period of coverage from the date of your retirement to the date of your election.
- Any regularly scheduled monthly premium that becomes due between your election and the end of the 45-day period.

Once the Plan Administrator receives this election form, you will be notified of the amount of the initial premium you must pay. **If you fail to pay the initial premium, or any subsequent monthly premium, in a timely fashion, your coverage will terminate. Checks are to be made payable to _____.**

Premium payments are generally due within 30 days after the first day of each month of coverage. Premium amounts change from time to time. You will be notified of any change in the premium amount.

The current monthly cost for COBRA continuation coverage under the plan for the different levels of coverage is as follows:

Individual $ _____

Husband & Wife $ _____

Parent & Child $ _____

Family $ _____

You currently have _____ coverage under the Company's plan. You are eligible to continue the coverage you had immediately prior to the occurrence of the qualifying event and unless you expressly elect otherwise, that coverage will be continued for you and any of the qualified beneficiaries covered by the plan as well. **If premium payment is not received on time, coverage will terminate and may not be reinstated.**

COBRA Coverage Election Agreement

I have read this form and the notice of my election rights. I understand my rights to elect continuation coverage and would like to take the action indicated below. I understand that if I elect continuation coverage and I fail to pay any premium payment on time, this coverage will terminate. I also agree to notify the Plan Administrator if I or any member of my family become(s) covered under another group health plan or entitled to Medicare.

Please check ONE option only.

☐ **I elect to continue individual coverage under the plan.**

Name   Relationship   Birth Date   Soc. Sec. #

_____

☐ **I elect to continue husband/wife coverage under the plan.**

Name   Relationship   Birth Date   Soc. Sec. #

_____

_____

☐ **I elect to continue parent/child coverage under the plan.**

Name   Relationship   Birth Date   Soc. Sec. #

_____

_____

☐ **I elect to continue family coverage under the plan. (Only to be checked by those qualified beneficiaries who had family coverage before the qualifying event, or in the event that a dependent child is born to or placed for adoption with a covered employee).**

List dependents to be covered:

Name   Relationship   Birth Date   Soc. Sec. #

_____

_____

_____

_____

☐ **I have read this form and the notice of rights. I am waiving my right to continuation coverage under the plan. (One form for each qualified beneficiary waiving his/her right to continuation coverage must be completed and forwarded to the plan administrator).**

Signature:_____ Date:_____

Name (Please Print):_____

Address: _____

Telephone: _____

Received By Plan Administrator:_____

Date: _____

If you elect to continue coverage, please return this signed election form to:

Name of Administrator: _____

Address: _____

Phone Number: _____

Your first monthly payment for continuation coverage must be enclosed with this notice or submitted within 45 days of your election pursuant to this notice. Subsequent monthly payments should be received by us the 1st day of each month. There is a grace period of at least 30 days for payment of a regularly scheduled premium.

IMPORTANT

The group insurance Plan under which coverage for you and, if applicable, your covered dependents, if any, is being continued also contains a conversion privilege which permits applying for an individual policy of medical care benefits. If you elect to continue coverage under the group Plan, application for such an individual policy may also be made upon cessation of the continued coverage, during or at the end of the applicable continuation period. If you do not elect to continue coverage under the group insurance plan you may also apply for an individual policy. If you are interested in making such a conversion, please contact your group Plan Administrator for more details.

Sample Hiring Record Form

Position: _____ Position Control Number: _____

Application Dates: Opened _____ Closed _____

Recruiting Sources Used: _____

_____

_____

Date	Applicants	Sex	White	Black	His-panic	Asian	Am. Indian
	M						
Totals F							

Applicant Selected _____

Sex: _____ Ethnicity: _____

Date: _____

Sample Employment Application

Position applied for_____ Date _____

Type of employment desired: Full-time_____   Part-time_____

Personal Information

Name _____ Social Security Number_____

Address_____

Phone _____ Date available

Employment History (list most recent employers/positions first)

Employer_____ Address_____ Phone _____

Position Title_____ Employed from (mo./yr.)_____ To (mo./yr.)_____

Duties_____

Supervisor_____ Last wage/salary: $_____ per _____

Reason for leaving_____

Employer_____ Address_____ Phone _____

Position Title_____ Employed from (mo./yr.)_____ To (mo./yr.)_____

Duties_____

Supervisor_____ Last wage/salary: $_____ per _____

Reason for leaving_____

Employer_____ Address_____ Phone _____

Position Title_____ Employed from (mo./yr.)_____ To (mo./yr.)_____

Duties_____

Supervisor_____ Last wage/salary: $_____ per _____

Reason for leaving_____

Employer_____ Address_____ Phone _____

Position Title_____ Employed from (mo./yr.)_____ To (mo./yr.)_____

Duties_____

Supervisor_____ Last wage/salary: $_____ per _____

Reason for leaving_____

Employer_____ Address_____ Phone _____

Position Title_____ Employed from (mo./yr.)_____ To (mo./yr.)_____

Duties_____

Supervisor_____ Last wage/salary: $_____ per _____

Reason for leaving_____

Skills

List any special skills you have. In addition, list special equipment you can operate: _____

_____

_____

_____

Education/Training Background

High School Graduate   ☐ Yes   ☐ No   ☐ GED

School _____   Date graduated _____

Trade school or certification   ☐ Yes   ☐ No   Trade studied/Certification:

School _____   _____

College or university   ☐ Yes   ☐ No   Degree _____

School _____   Dates attended _____

Qualifications

Are you legally eligible for employment in the United States?   ☐ Yes   ☐ No

Are you able to perform the essential functions of the position applied for with or without reasonable accommodations?   ☐ Yes   ☐ No

If hired, can you furnish proof that you are 18 years of age or older, or if under 18, do you have a permit to work?   ☐ Yes   ☐ No

If No, explain:

Have you been convicted of a felony or released from prison in the past five years?
☐ Yes   ☐ No

Note: A "Yes" answer does not automatically disqualify you from employment since the nature of the offense, date, and type of job for which you are applying will be considered. You may omit information regarding convictions that have been expunged.

I certify that the information contained in this application is correct. I understand that the misrepresentation or omission of information called for in this application is cause for refusal to hire or, if hired, is cause for immediate dismissal. I authorize the investigation of all statements contained in this application and authorize any of the persons or organizations referenced in this application to furnish [YOUR COMPANY] any and all information concerning my previous employment, education, or any other information they might have, personal or otherwise, with regard to any of the subjects covered by this application and release all such parties from any and all liability or damages that may result from their furnishing such information to [YOUR COMPANY].

Further, I understand and agree that my employment is at-will and for no definite period and may, regardless of the date of payment of my wages and salary, be terminated at any time with or without notice at the option of either [YOUR COMPANY] or myself. Further, I understand that no officer, agent, representative, or employee of [YOUR COMPANY] has any authority to enter into any agreement for employment for any specified period of time or to make any agreement contrary to that contained in the previous sentence.

I also authorize [YOUR COMPANY] to deduct from my wages any amounts that may be due it as a result of overpayment of wages, loss or destruction of its property, or any other amounts which I may lawfully owe [YOUR COMPANY] or for which I have received full consideration. In the event I become an employee of [YOUR COMPANY], I agree to comply with all rules and regulations and understand that the rules and regulations may be changed, interpreted, withdrawn, or added to by [YOUR COMPANY] at any time at its sole option and without any prior notice and that I may be terminated or disciplined for any violations.

Signature _____ Date _____

[Check your state laws and regulations for any additional notices or disclaimers that must be included and explained in the Employment Application Form. For example:

UNDER MARYLAND LAW AN EMPLOYER MAY NOT REQUIRE OR DEMAND ANY APPLICANT FOR EMPLOYMENT OR PROSPECTIVE EMPLOYMENT OR ANY EMPLOYEE TO SUBMIT TO OR TAKE A POLY-GRAPH, LIE DETECTOR, OR SIMILAR TEST OR EXAMINATION AS A CONDITION OF EMPLOYMENT OR CONTINUED EMPLOYMENT. ANY EMPLOYER WHO VIOLATES THIS PROVISION IS GUILTY OF A MIS-DEMEANOR AND SUBJECT TO A FINE NOT TO EXCEED $100.00.]

Signature _____ Date _____

Note: This application will remain active for seven days from the date of filing. The applicant must file a new application to be considered for job openings occurring subsequent to the seven day period.

An Equal Opportunity Employer

Sample Applicant Data Record

Applicants are considered for all positions, and employees are treated during employment without regard to race, color, religion, sex, national origin, age, marital or veteran status, medical condition, or handicap.

As employers/government contractors, we comply with government regulations and affirmative action responsibilities.

Solely to help us comply with government record keeping, reporting and other legal requirements, please fill out the Applicant Data Record. We appreciate your cooperation.

The information provided on this form is for periodic government reporting and will be kept in a Confidential File separate from the Employment Application.

Please Print   Date _____

Position applied for_____

Referral source: ☐ Advertisement ☐ Internet ☐ Friend ☐ Relative ☐ Walk-in ☐ Employment agency

☐ Other_____

Name_____ Phone_____

Address_____

Affirmative Action Survey

Government agencies require periodic reports on the sex, ethnicity, handicapped, and veteran status of applicants. This data is for analysis and affirmative action only. Submission of information is voluntary.

Check one   ☐ Male   ☐ Female

Check one: ☐ White   ☐ Black   ☐ Hispanic
               ☐ Asian/Pacific Islander   ☐ American Indian/Alaskan Native

Check if any of the following are applicable:

☐ Vietnam-era veteran   ☐ Disabled veteran   ☐ Individual with a disability

Employment/Education Reference Verification Form

YOUR COMPANY
Address
City, State, zip code
Phone; E-mail
Date

To _____

_____

_____

The person named below has applied for a position with [YOUR COMPANY]. You were named as a reference by that person. Please complete the relevant information below and return this form to us at the above address. Thank you.

Applicant's name_____ Social Security Number_____

☐ **Previous Employer** (Please verify, correct, or supply the information below)

1. Applicant's position with you_____ ☐ Yes ☐ No

2. Dates of employment: From (mo./yr.)_____ To (mo./yr.)_____ ☐ Yes ☐ No

3. Main job duties _____ ☐ Yes ☐ No

4. Reason for separation ☐ Resigned ☐ Dismissed ☐ Laid off ☐ Other

5. Please rate this person on attendance ☐ Acceptable ☐ Unacceptable

6. Please rate this person on work performance ☐ Acceptable ☐ Unacceptable

7. Would you reemploy this person? ☐ Yes ☐ No If "No," why not?_____

_____

8. Completed by_____
              Your Name   Position   Date

☐ **Educational or Training Institution** (Please verify, correct, or supply the information below)

The person identified above states that he or she attended your institution:

1. From (mo./yr.) _____ To (mo./yr.) _____ ☐ Yes ☐ No

2. Graduation date _____ ☐ Yes ☐ No ☐ Does not apply

3. Degree/certificate/license _____ ☐ Yes ☐ No

4. Completed by _____
              Your Name   Position   Date

I authorize the employer/educational institution noted above to supply the information requested. I release respondent and the organization from liability in answering the items correctly.

Applicant's signature_____ Date _____

Internal use—Phone check: Date_____ By _____

Results: _____

Sample Offer of Employment Letter

Dear_____

I am pleased to confirm the offer of employment made to you by [YOUR COMPANY]. The details of the offer are listed below:

1. Position

2. Starting wage of $_____ per hour/year

3. Starting day_____ starting time_____

4. On your first day, report to_____ Phone: _____

Location_____

[YOUR COMPANY], in compliance with the 1986 Immigration Reform and Control Act, hires and retains in employment only persons who legally reside in the U.S. and have valid work authorization for U.S. employment. In accordance with the law's verification requirements, all applicants offered a job must provide proof of work eligibility and U.S. citizenship or lawful residence. The documents required are described in an attachment to this letter. If you do not think you can produce these documents for our verification when you report to duty, please let us know as soon as possible and we will try to assist you. Unfortunately, if you do not have the documents, the law prohibits us from hiring you.

5. Please remember that your continued employment with us depends on the successful completion and receipt of any of the item noted below:

☐ School transcripts or certificates
☐ Reference checks
☐ Medical examination

☐ Other_____

We are pleased that you will be joining us and look forward to a mutually rewarding relationship. However, please keep in mind that this letter is not a contract of employment, and all employees at [YOUR COMPANY] work on an at-will basis. This employment relationship can be dissolved at any time by either party.

We look forward to working with you.

Sincerely yours,

Official Signature

enclosures

Offer of Employment Letter Attachment

The Immigration Reform and Control Act of 1986

[YOUR COMPANY], in compliance with the 1986 Immigration Reform and Control Act, hires and retains in employment only persons who legally reside in the United States and have valid work authorization for U.S. employment. In accordance with the law's verification requirements, all applicants offered a job must provide proof of work eligibility and U.S. citizenship or lawful residence. The documents required are described below. Unfortunately, if you do not have the documents, the law prohibits us from hiring you.

Employment Authorization and Identity Verification Requirements

An employer complies with the verification requirements of the law if the individual to be hired presents one of the following documents (primary verification test):

- A United States passport
- A Certificate of United States citizenship
- A U.S. Citizen Identification Card
- A certificate of United States naturalization
- An unexpired foreign passport with:

   1. An unexpired stamp which reads "Process for I-551 . . ."
   2. A Form 1-94 in the same name with an unexpired employment authorization stamp; proposed employment should not conflict with restrictions or limitations identified on Form 1-94.

- An Alien Registration Card ("green card") with a photo of the bearer
- An unexpired Work Permit, issued by U.S. Immigration and Naturalization Service
- An unexpired Temporary Resident Card
- An unexpired Reentry Permit

If you cannot produce one of the documents listed above, you may still be hired if you can produce one document demonstrating your employment eligibility and another demonstrating your identity verification. Documents that demonstrate employment eligibility consist of the following:

- A social security account number card (other than a card which states that it does not authorize employment in the United States)
- A U.S. birth or nationality certificate
- Report of the U.S. Citizen Birth Abroad, issued by U.S. Department of State
- Form 1-94 with employment authorization stamp
- Native American tribal document

Documents that that can be used to establish identity including the following:

- A driver's license or similar document issued for the purpose of identification by a state, *provided* that it contains a photograph and personal identifying information;
- An original identity issued by any state for purpose of identification that either:

   1. Bears a photo of the individual, or
   2. Contains personal identifying information

- In the case of individuals residing in a state which does not issue an identification document (other than a driver's license), the following:

   1. Notice of discharge from U.S. Forces
   2. Document evidencing active duty of reserve status in the U.S. Armed Forces
   3. School ID card with a photograph
   4. U.S. military card

These appropriate documents must be examined prior to hiring any individual after January 1, 1987.

U.S. Department of Justice
Immigration and Naturalization Service

OMB No. 1115-0136

Employment Eligibility Verification

INSTRUCTIONS
PLEASE READ ALL INSTRUCTIONS CAREFULLY BEFORE COMPLETING THIS FORM.

Anti-Discrimination Notice. It is illegal to discriminate against any individual (other than an alien not authorized to work in the U.S.) in hiring, discharging, or recruiting or referring for a fee because of that individual's national origin or citizenship status. It is illegal to discriminate against work eligible individuals. Employers **CANNOT** specify which document(s) they will accept from an employee. The refusal to hire an individual because of a future expiration date may also constitute illegal discrimination.

Section 1 - Employee.
All employees, citizens and noncitizens, hired after November 6, 1986, must complete Section 1 of this form at the time of hire, which is the actual beginning of employment.**The employer is responsible for ensuring that Section 1 is timely and properly completed.**

Preparer/Translator Certification The Preparer/Translator Certification must be completed if Section 1 is prepared by a person other than the employee. A preparer/translator may be used only when the employee is unable to complete Section 1 on his/her own. However, the employee must still sign Section 1.

Section 2 - Employer.
For the purpose of completing this form, the term "employer" includes those recruiters and referrers for a fee who are agricultural associations, agricultural employers or farm labor contractors.

Employers must complete Section 2 by examining evidence of identity and employment eligibility within three (3) business days of the date employment begins. If employees are authorized to work, but are unable to present the required document(s) within three business days, they must present a receipt for the application of the document(s) within three business days and the actual document(s) within ninety (90) days. However, if employers hire individuals for a duration of less than three business days, Section 2 must be completed at the time employment begins.**Employers must record: 1)** document title; **2)** issuing authority; **3)** document number, **4)** expiration date, if any; and **5)** the date employment begins. Employers must sign and date the certification. Employees must present original documents. Employers may, but are not required to, photocopy the document(s) presented. These photocopies may only be used for the verification process and must be retained with the I-9. **However, employers are still responsible for completing the I-9.**

Section 3 - Updating and Reverification.
Employers must complete Section 3 when updating and/or reverifying the I-9. Employers must reverify employment eligibility of their employees on or before the expiration date recorded in Section 1. Employers **CANNOT** specify which document(s) they will accept from an employee.

- If an employee's name has changed at the time this form is being updated/ reverified, complete Block A.

- If an employee is rehired within three (3) years of the date this form was originally completed and the employee is still eligible to be employed on the same basis as previously indicated on this form (updating), complete Block B and the signature block.

- If an employee is rehired within three (3) years of the date this form was originally completed and the employee's work authorization has expired **or** if a current employee's work authorization is about to expire (reverification), complete Block B and:
  - examine any document that reflects that the employee is authorized to work in the U.S. (see List A **or** C),
  - record the document title, document number and expiration date (if any) in Block C, and complete the signature block.

Photocopying and Retaining Form I-9. A blank I-9 may be reproduced, provided both sides are copied. The Instructions must be available to all employees completing this form. Employers must retain completed I-9s for three (3) years after the date of hire or one (1) year after the date employment ends, whichever is later.

For more detailed information, you may refer to the INS Handbook for Employers, (Form M-274). You may obtain the handbook at your local INS office.

Privacy Act Notice. The authority for collecting this information is the Immigration Reform and Control Act of 1986, Pub. L. 99-603 (8 USC 1324a)

This information is for employers to verify the eligibility of individuals for employment to preclude the unlawful hiring, or recruiting or referring for a fee, of aliens who are not authorized to work in the United States.

This information will be used by employers as a record of their basis for determining eligibility of an employee to work in the United States. The form will be kept by the employer and made available for inspection by officials of the U.S. Immigration and Naturalization Service, the Department of Labor and the Office of Special Counsel for Immigration Related Unfair Employment Practices.

Submission of the information required in this form is voluntary. However, an individual may not begin employment unless this form is completed, since employers are subject to civil or criminal penalties if they do not comply with the Immigration Reform and Control Act of 1986.

Reporting Burden. We try to create forms and instructions that are accurate, can be easily understood and which impose the least possible burden on you to provide us with information. Often this is difficult because some immigration laws are very complex. Accordingly, the reporting burden for this collection of information is computed as follows: **1)** learning about this form, 5 minutes; **2)** completing the form, 5 minutes; and **3)** assembling and filing (recordkeeping) the form, 5 minutes, for an average of 15 minutes per response. If you have comments regarding the accuracy of this burden estimate, or suggestions for making this form simpler, you can write to the Immigration and Naturalization Service, HQPDI, 425 I Street, N.W., Room 4034, Washington, DC 20536. OMB No. 1115-0136.

EMPLOYERS MUST RETAIN COMPLETED FORM I-9
PLEASE DO NOT MAIL COMPLETED FORM I-9 TO INS

Form I-9 (Rev. 11-21-91)N

U.S. Department of Justice
Immigration and Naturalization Service

OMB No. 1115-0136

Employment Eligibility Verification

Please read instructions carefully before completing this form. The instructions must be available during completion of this form. ANTI-DISCRIMINATION NOTICE: It is illegal to discriminate against work eligible individuals. Employers CANNOT specify which document(s) they will accept from an employee. The refusal to hire an individual because of a future expiration date may also constitute illegal discrimination.

Section 1. Employee Information and Verification. To be completed and signed by employee at the time employment begins.

Print Name: Last	First	Middle Initial	Maiden Name

Address *(Street Name and Number)*		Apt.#	Date of Birth *(month/day/year)*

City	State	Zip Code	Social Security #

I am aware that federal law provides for imprisonment and/or fines for false statements or use of false documents in connection with the completion of this form.

I attest, under penalty of perjury, that I am (check one of the following):
- ☐ A citizen or national of the United States
- ☐ A Lawful Permanent Resident (Alien # A_____)
- ☐ An alien authorized to work until ___/___/___

(Alien # or Admission #)_____

Employee's Signature	Date *(month/day/year)*

Preparer and/or Translator Certification. *(To be completed and signed if Section 1 is prepared by a person other than the employee.) I attest, under penalty of perjury, that I have assisted in the completion of this form and that to he best of my knowledge the information is true and correct.*

Preparer's/Translator's Signature	Print Name

Address *(Street Name and Number, City, State, Zip Code)*	Date *(month/day/year)*

Section 2. Employer Review and Verification. To be completed and signed by employer. Examine one document from List A OR examine one document from List B and one from List C, as listed on the reverse of this form, and record the title, number and expiration date, if any, of the document(s)

List A	OR	List B	AND	List C
Document title:_____		_____		_____
Issuing authority:_____		_____		_____
Document #: _____		_____		_____
Expiration Date *(if any):* ___/___/___		___/___/___		___/___/___
Document #: _____				
Expiration Date *(if any):* ___/___/___				

CERTIFICATION - I attest, under penalty of perjury, that I have examined the document(s) presented by the above-named employee, that the above-listed document(s) appear to be genuine and to relate to the employee named, that the employee began employment on *(month/day/year)* ___/___/___ **and that to the best of my knowledge the employee is eligible to work in the United States. (State employment agencies may omit the date the employee began employment.)**

Signature of Employer or Authorized Representative	Print Name	Title

Business or Organization Name	Address *(Street Name and Number, City, State, Zip Code)*	Date *(month/day/year)*

Section 3. Updating and Reverification. To be completed and signed by employer.

A. New Name *(if applicable)*	B. Date of rehire *(month/day/year) (if applicable)*

C. If employee's previous grant of work authorization has expired, provide the information below for the document that establishes current employment eligibility.

Document Title:_____ Document #: _____ Expiration Date (if any):___/___/___

I attest, under penalty of perjury, that to the best of my knowledge, this employee is eligible to work in the United States, and if the employee presented document(s), the document(s) I have examined appear to be genuine and to relate to the individual.

Signature of Employer or Authorized Representative	Date *(month/day/year)*

Form I-9 (Rev. 11-21-91)N Page 2

LISTS OF ACCEPTABLE DOCUMENTS

LIST A		LIST B		LIST C
Documents that Establish Both Identity and Employment Eligibility	**OR**	**Documents that Establish Identity**	**AND**	**Documents that Establish Employment Eligibility**

LIST A — Documents that Establish Both Identity and Employment Eligibility

1. U.S. Passport (unexpired or expired)

2. Certificate of U.S. Citizenship *(INS Form N-560 or N-561)*

3. Certificate of Naturalization *(INS Form N-550 or N-570)*

4. Unexpired foreign passport, with *I-551 stamp or* attached INS Form I-94 indicating unexpired employment authorization

5. Permanent Resident Card or Alien Registration Receipt Card with photograph *(INS Form I-151 or I-551)*

6. Unexpired Temporary Resident Card *(INS Form I-688)*

7. Unexpired Employment Authorization Card *(INS Form I-688A)*

8. Unexpired Reentry Permit *(INS Form I-327)*

9. Unexpired Refugee Travel Document *(INS Form I-571)*

10. Unexpired Employment Authorization Document issued by the INS which contains a photograph *(INS Form I-688B)*

OR

LIST B — Documents that Establish Identity

1. Driver's license or ID card issued by a state or outlying possession of the United States provided it contains a photograph or information such as name, date of birth, gender, height, eye color and address

2. ID card issued by federal, state or local government agencies or entities, provided it contains a photograph or information such as name, date of birth, gender, height, eye color and address

3. School ID card with a photograph

4. Voter's registration card

5. U.S. Military card or draft record

6. Military dependent's ID card

7. U.S. Coast Guard Merchant Mariner Card

8. Native American tribal document

9. Driver's license issued by a Canadian government authority

For persons under age 18 who are unable to present a document listed above:

10. School record or report card

11. Clinic, doctor or hospital record

12. Day-care or nursery school record

AND

LIST C — Documents that Establish Employment Eligibility

1. U.S. social security card issued by the Social Security Administration *(other than a card stating it is not valid for employment)*

2. Certification of Birth Abroad issued by the Department of State *(Form FS-545 or Form DS-1350)*

3. Original or certified copy of a birth certificate issued by a state, county, municipal authority or outlying possession of the United States bearing an official seal

4. Native American tribal document

5. U.S. Citizen ID Card *(INS Form I-197)*

6. ID Card for use of Resident Citizen in the United States *(INS Form I-179)*

7. Unexpired employment authorization document issued by the INS *(other than those listed under List A)*

Illustrations of many of these documents appear in Part 8 of the Handbook for Employers (M-274)

New Employee Orientation Procedures Checklist

Put the new employee's name at the top of the sheet, check off items as they are completed, and retain this sheet in the employee's personnel folder.

Employee_____ Hire Date_____

Have the employee complete the following:

☐ The employment application form (if not done already)
☐ A list of emergency contacts
☐ The 1-9 form
☐ State and federal tax withholding information
☐ Other payroll information (as appropriate)
☐ Any applicable benefits registration forms

Provide the employee with a copy of the employee handbook and review it with him or her.

☐ Acknowledgement of Receipt of Employee Handbook (make sure you have a signed copy of this form, which appears at the end of Section II)

You should be sure to:

☐ Add the employee to the payroll system
☐ Notify other employees and introduce the new employee
☐ Perform proper safety training

Completed by_____ Date_____

OSHA
Forms for Recording
Work-Related Injuries and Illnesses

What's Inside...

In this package, you'll find everything you need to complete OSHA's *Log* and the *Summary of Work-Related Injuries and Illnesses* for the next several years. On the following pages, you'll find:

▼ **An Overview: Recording Work-Related Injuries and Illnesses** — General instructions for filling out the forms in this package and definitions of terms you should use when you classify your cases as injuries or illnesses.

▼ **How to Fill Out the Log** — An example to guide you in filling out the *Log* properly.

▼ **Log of Work-Related Injuries and Illnesses** — Several pages of the *Log* (but you may make as many copies of the *Log* as you need.) Notice that the *Log* is separate from the *Summary*.

▼ **Summary of Work-Related Injuries and Illnesses** — Removable *Summary* pages for easy posting at the end of the year. Note that you post the *Summary* only, not the *Log*.

▼ **Worksheet to Help You Fill Out the Summary** — A worksheet for figuring the average number of employees who worked for your establishment and the total number of hours worked.

▼ **OSHA's 301: Injury and Illness Incident Report** — Several copies of the OSHA 301 to provide details about the incident. You may make as many copies as you need or use an equivalent form.

Take a few minutes to review this package. If you have any questions, *visit us online at www.osha. gov* or *call your local OSHA office.* We'll be happy to help you.

An Overview:
Recording Work-Related Injuries and Illnesses

The Occupational Safety and Health (OSH) Act of 1970 requires certain employers to prepare and maintain records of work-related injuries and illnesses. Use these definitions when you classify cases on the Log. OSHA's recordkeeping regulation (see 29 CFR Part 1904) provides more information about the definitions below.

The *Log of Work-Related Injuries and Illnesses* (Form 300) is used to classify work-related injuries and illnesses and to note the extent and severity of each case. When an incident occurs, use the *Log* to record specific details about what happened and how it happened. The *Summary* — a separate form (Form 300A) — shows the totals for the year in each category. At the end of the year, post the *Summary* in a visible location so that your employees are aware of the injuries and illnesses occurring in their workplace.

Employers must keep a *Log* for each establishment or site. If you have more than one establishment, you must keep a separate *Log* and *Summary* for each physical location that is expected to be in operation for one year or longer.

Note that your employees have the right to review your injury and illness records. For more information, see 29 Code of Federal Regulations Part 1904.35, *Employee Involvement*.

Cases listed on the *Log of Work-Related Injuries and Illnesses* are not necessarily eligible for workers' compensation or other insurance benefits. Listing a case on the *Log* does not mean that the employer or worker was at fault or that an OSHA standard was violated.

When is an injury or illness considered work-related?

An injury or illness is considered work-related if an event or exposure in the work environment caused or contributed to the condition or significantly aggravated a preexisting condition. Work-relatedness is presumed for injuries and illnesses resulting from events or exposures occurring in the workplace, unless an exception specifically applies. See 29 CFR Part 1904.5(b)(2) for the exceptions. The work environment includes the establishment and other locations where one or more employees are working or are present as a condition of their employment. See 29 CFR Part 1904.5(b)(1).

Which work-related injuries and illnesses should you record?

Record those work-related injuries and illnesses that result in:

▼ death,
▼ loss of consciousness,
▼ days away from work,
▼ restricted work activity or job transfer, or
▼ medical treatment beyond first aid.

You must also record work-related injuries and illnesses that are significant (as defined below) or meet any of the additional criteria listed below.

You must record any significant work-related injury or illness that is diagnosed by a physician or other licensed health care professional. You must record any work-related case involving cancer, chronic irreversible disease, a fractured or cracked bone, or a punctured eardrum. See 29 CFR 1904.7.

What are the additional criteria?

You must record the following conditions when they are work-related:

▼ any needlestick injury or cut from a sharp object that is contaminated with another person's blood or other potentially infectious material;
▼ any case requiring an employee to be medically removed under the requirements of an OSHA health standard;
▼ tuberculosis infection as evidenced by a positive skin test or diagnosis by a physician or other licensed health care professional after exposure to a known case of active tuberculosis.

What is medical treatment?

Medical treatment includes managing and caring for a patient for the purpose of combating disease or disorder. The following are not considered medical treatments and are NOT recordable:

▼ visits to a doctor or health care professional solely for observation or counseling;
▼ diagnostic procedures, including administering prescription medications that are used solely for diagnostic purposes; and
▼ any procedure that can be labeled first aid. *(See below for more information about first aid.)*

What do you need to do?

1. Within 7 calendar days after you receive information about a case, decide if the case is recordable under the OSHA recordkeeping requirements.

2. Determine whether the incident is a new case or a recurrence of an existing one.

3. Establish whether the case was work-related.

4. If the case is recordable, decide which form you will fill out as the injury and illness incident report.

   You may use *OSHA's 301: Injury and Illness Incident Report* or an equivalent form. Some state workers compensation, insurance, or other reports may be acceptable substitutes, as long as they provide the same information as the OSHA 301.

How to work with the Log

1. Identify the employee involved unless it is a privacy concern case as described below.

2. Identify when and where the case occurred.

3. Describe the case, as specifically as you can.

4. Classify the seriousness of the case by recording the **most serious outcome** associated with the case, with column J (Other recordable cases) being the least serious and column G (Death) being the most serious.

5. Identify whether the case is an injury or illness. If the case is an injury, check the injury category. If the case is an illness, check the appropriate illness category.

What is first aid?

If the incident required only the following types of treatment, consider it first aid. Do NOT record the case if it involves only:

▶ using non-prescription medications at non-prescription strength;

▶ administering tetanus immunizations;

▶ cleaning, flushing, or soaking wounds on the skin surface;

▶ using wound coverings, such as bandages, BandAids™, gauze pads, etc., or using SteriStrips™ or butterfly bandages.

▶ using hot or cold therapy;

▶ using any totally non-rigid means of support, such as elastic bandages, wraps, non-rigid back belts, etc.;

▶ using temporary immobilization devices while transporting an accident victim (splints, slings, neck collars, or back boards).

▶ drilling a fingernail or toenail to relieve pressure, or draining fluids from blisters;

▶ using eye patches;

▶ using simple irrigation or a cotton swab to remove foreign bodies not embedded in or adhered to the eye;

▶ using irrigation, tweezers, cotton swab or other simple means to remove splinters or foreign material from areas other than the eye;

▶ using finger guards;

▶ using massages;

▶ drinking fluids to relieve heat stress

How do you decide if the case involved restricted work?

Restricted work activity occurs when, as the result of a work-related injury or illness, an employer or health care professional keeps, or recommends keeping, an employee from doing the routine functions of his or her job or from working the full workday that the employee would have been scheduled to work before the injury or illness occurred.

How do you count the number of days of restricted work activity or the number of days away from work?

Count the number of calendar days the employee was on restricted work activity or was away from work as a result of the recordable injury or illness. Do not count the day on which the injury or illness occurred in this number. Begin counting days from the day after the incident occurs. If a single injury or illness involved both days away from work and days of restricted work activity, enter the total number of days for each. You may stop counting days of restricted work activity or days away from work once the total of either or the combination of both reaches 180 days.

Under what circumstances should you NOT enter the employee's name on the OSHA Form 300?

You must consider the following types of injuries or illnesses to be privacy concern cases:

▶ an injury or illness to an intimate body part or to the reproductive system,

▶ an injury or illness resulting from a sexual assault,

▶ a mental illness,

▶ a case of HIV infection, hepatitis, or tuberculosis,

▶ a needlestick injury or cut from a sharp object that is contaminated with blood or other potentially infectious material (see 29 CFR Part 1904.8 for definition), and

▶ other illnesses, if the employee independently and voluntarily requests that his or her name not be entered on the log.

You must not enter the employee's name on the OSHA 300 *Log* for these cases. Instead, enter "privacy case" in the space normally used for the employee's name. You must keep a separate, confidential list of the case numbers and employee names for the establishment's privacy concern cases so that you can update the cases and provide information to the government if asked to do so.

If you have a reasonable basis to believe that information describing the privacy concern case may be personally identifiable even though the employee's name has been omitted, you may use discretion in describing the injury or illness on both the OSHA 300 and 301 forms. You must enter enough information to identify the cause of the incident and the general severity of the injury or illness, but you do not need to include details of an intimate or private nature.

What if the outcome changes after you record the case?

If the outcome or extent of an injury or illness changes after you have recorded the case, simply draw a line through the original entry or, if you wish, delete or white-out the original entry. Then write the new entry where it belongs. Remember, you need to record the most serious outcome for each case.

Classifying injuries

An injury is any wound or damage to the body resulting from an event in the work environment.

Examples: Cut, puncture, laceration, abrasion, fracture, bruise, contusion, chipped tooth, amputation, insect bite, electrocution, or a thermal, chemical, electrical, or radiation burn. Sprain and strain injuries to muscles, joints, and connective tissues are classified as injuries when they result from a slip, trip, fall or other similar accidents.

Optional

Calculating Injury and Illness Incidence Rates

What is an incidence rate?

An incidence rate is the number of recordable injuries and illnesses occurring among a given number of full-time workers (usually 100 full-time workers) over a given period of time (usually one year). To evaluate your firm's injury and illness experience over time or to compare your firm's experience with that of your industry as a whole, you need to compute your incidence rate. Because a specific number of workers and a specific period of time are involved, these rates can help you identify problems in your workplace and/or progress you may have made in preventing work-related injuries and illnesses.

How do you calculate an incidence rate?

You can compute an occupational injury and illness incidence rate for all recordable cases or for cases that involved days away from work for your firm quickly and easily. The formula requires that you follow instructions in paragraph (a) below for the total recordable cases or those in paragraph (b) for cases that involved days away from work, *and* for both rates the instructions in paragraph (c).

(a) *To find out the total number of recordable injuries and illnesses that occurred during the year,* count the number of line entries on your OSHA Form 300, or refer to the OSHA Form 300A and sum the entries for columns (G), (H), (I), and (J).

(b) *To find out the number of injuries and illnesses that involved days away from work,* count the number of line entries on your OSHA Form 300 that received a check mark in column (H), or refer to the entry for column (H) on the OSHA Form 300A.

(c) *The number of hours all employees actually worked during the year.* Refer to OSHA Form 300A and optional worksheet to calculate this number.

You can compute the incidence rate for all recordable cases of injuries and illnesses using the following formula:

Total number of injuries and illnesses ÷ Number of hours worked by all employees × 200,000 hours = Total recordable case rate

(The 200,000 figure in the formula represents the number of hours 100 employees working 40 hours per week, 50 weeks per year would work, and provides the standard base for calculating incidence rates.)

You can compute the incidence rate for recordable cases involving days away from work, days of restricted work activity or job transfer (DART) using the following formula:

(Number of entries in column H + Number of entries in column I) ÷ Number of hours worked by all employees × 200,000 hours = DART incidence rate

You can use the same formula to calculate incidence rates for other variables such as cases involving restricted work activity (column I) on Form 300A), cases involving skin disorders (column (M-2) on Form 300A), etc. Just substitute the appropriate total for these cases, from Form 300A, into the formula in place of the total number of injuries and illnesses.

What can I compare my incidence rate to?

The Bureau of Labor Statistics (BLS) conducts a survey of occupational injuries and illnesses each year and publishes incidence rate data by various classifications (e.g., by industry, by employer size, etc.). You can obtain these published data at www.bls.gov or by calling a BLS Regional Office.

Worksheet

Total number of recordable injuries and illnesses in your establishment

[_____]

÷

[_____]

Hours worked by all your employees

X 200,000 =

[_____]

Total recordable cases incidence rate

Total number of recordable injuries and illnesses with a checkmark in column H or column I

[_____]

÷

[_____]

Hours worked by all your employees

X 200,000 =

[_____]

DART incidence rate

Optional

Calculating Injury and Illness Incidence Rates

What is an incidence rate?

An incidence rate is the number of recordable injuries and illnesses occurring among a given number of full-time workers (usually 100 full-time workers) over a given period of time (usually one year). To evaluate your firm's injury and illness experience over time or to compare your firm's experience with that of your industry as a whole, you need to compute your incidence rate. Because a specific number of workers and a specific period of time are involved, these rates can help you identify problems in your workplace and/or progress you may have made in preventing work-related injuries and illnesses.

How do you calculate an incidence rate?

You can compute an occupational injury and illness incidence rate for all recordable cases or for cases that involved days away from work for your firm quickly and easily. The formula requires that you follow instructions in paragraph (a) below for the total recordable cases or those in paragraph (b) for cases that involved days away from work, *and* for both rates the instructions in paragraph (c).

(a) *To find out the total number of recordable injuries and illnesses that occurred during the year,* count the number of line entries on your OSHA Form 300, or refer to the OSHA Form 300A and sum the entries for columns (G), (H), (I), and (J).

(b) *To find out the number of injuries and illnesses that involved days away from work, count* the number of line entries on your OSHA Form 300 that received a check mark in column (H), or refer to the entry for column (H) on the OSHA Form 300A.

(c) *The number of hours all employees actually worked during the year.* Refer to OSHA Form 300A and optional worksheet to calculate this number.

You can compute the incidence rate for all recordable cases of injuries and illnesses using the following formula:

Total number of injuries and illnesses ÷ Number of hours worked by all employees x 200,000 hours = Total recordable case rate

(The 200,000 figure in the formula represents the number of hours 100 employees working 40 hours per week, 50 weeks per year would work, and provides the standard base for calculating incidence rates.)

You can compute the incidence rate for recordable cases involving days away from work, days of restricted work activity or job transfer (DART) using the following formula:

(Number of entries in column H + Number of entries in column I) ÷ Number of hours worked by all employees x 200,000 hours = DART incidence rate

You can use the same formula to calculate incidence rates for other variables such as cases involving restricted work activity (column (I) on Form 300A), cases involving skin disorders (column (M-2) on Form 300A), etc. Just substitute the appropriate total for these cases, from Form 300A, into the formula in place of the total number of injuries and illnesses.

What can I compare my incidence rate to?

The Bureau of Labor Statistics (BLS) conducts a survey of occupational injuries and illnesses each year and publishes incidence rate data by various classifications (e.g., by industry, by employer size, etc.). You can obtain these published data at www.bls.gov or by calling a BLS Regional Office.

Worksheet

Total number of recordable injuries
and illnesses in your establishment

☐ ÷ ☐	X 200,000 =	☐	Total recordable cases incidence rate

Hours worked by all your employees

Total number of recordable injuries
and illnesses with a checkmark in
column H or column I

☐ ÷ ☐	X 200,000 =	☐	DART incidence rate

Hours worked by all your employees

How to Fill Out the Log

The *Log of Work-Related Injuries and Illnesses* is used to classify work-related injuries and illnesses and to note the extent and severity of each case. When an incident occurs, use the *Log* to record specific details about what happened and how it happened.

If your company has more than one establishment or site, you must keep separate records for each physical location that is expected to remain in operation for one year or longer.

We have given you several copies of the *Log* in this package. If you need more than we provided, you may photocopy and use as many as you need.

The *Summary* — a separate form — shows the work-related injury and illness totals for the year in each category. At the end of the year, count the number of incidents in each category and transfer the totals from the *Log* to the *Summary*. Then post the *Summary* in a visible location so that your employees are aware of injuries and illnesses occurring in their workplace.

You don't post the *Log*. You post only the *Summary* at the end of the year.

OSHA's Form 300

Log of Work-Related Injuries and Illnesses

Year 20___

U.S. Department of Labor
Occupational Safety and Health Administration
Form approved OMB no. 1218-0176

You must record information about every work-related death and about every work-related injury or illness that involves loss of consciousness, restricted work activity or job transfer, days away from work, or medical treatment beyond first aid. You must also record significant work-related injuries and illnesses that are diagnosed by a physician or licensed health care professional. You must also record work-related injuries and illnesses that meet any of the specific recording criteria listed in 29 CFR Part 1904.8 through 1904.12. Feel free to use two lines for a single case if you need to. You must complete an injury and illness Incident Report (OSHA Form 301) or equivalent form for each injury or illness recorded on this form. If you're not sure whether a case is recordable, call your local OSHA office for help.

> **Attention:** This form contains information relating to employee health and must be used in a manner that protects the confidentiality of employees to the extent possible while the information is being used for occupational safety and health purposes.

Establishment name: *XYZ Company*
City: *Anywhere*   State: *MA*

Identify the person / Describe the case

(A) Case no.	(B) Employee's name	(C) Job title (e.g., Welder)	(D) Date of injury or onset of illness	(E) Where the event occurred (e.g., Loading dock north end)	(F) Describe injury or illness, parts of body affected, and object/substance that directly injured or made person ill (e.g., Second degree burns on right forearm from acetylene torch)
1	Mark Bagin	Welder	5/25	basement	fracture, left arm and left leg, fell from ladder
2	Shana Alexander	Foundry man	7/2	pouring deck	poisoning from lead fumes
3	Sam Sander	Electrician	8/5	2nd floor storeroom	broken left foot, fell over box
4	Ralph Boccella	Laborer	9/17	packaging dept	Back strain lifting boxes
5	Jarrod Daniels	Machine opt.	10/23	production floor	dust in eye

Classify the case

Using these four categories, check ONLY the most serious result for each case:
(G) Death, (H) Days away from work, (I) Remained at work — Job transfer or restriction, (J) Other recordable cases.

Enter the number of days the injured or ill worker was:
(K) On job transfer or restriction — 12 days; 30 days; 7 days; ___ days; ___ days
(L) Away from work — 15 days; 30 days; ___ days; 3 days; ___ days

Check the "Injury" column or choose one type of illness:
(M) (1) Injury, (2) Skin disorder, (3) Respiratory condition, (4) Poisoning, (5) All other illnesses.

Be as specific as possible. You can use two lines if you need more room.

Revise the log if the injury or illness progresses and the outcome is more serious than you originally recorded for the case. Cross out, erase, or white-out the original entry.

Choose ONE of these categories. Classify the case by recording the most serious outcome of the case, with column J (Other recordable cases) being the least serious and column G (Death) being the most serious.

Note whether the case involves an injury or an illness.

OSHA's Form 300

Log of Work-Related Injuries and Illnesses

Attention: This form contains information relating to employee health and must be used in a manner that protects the confidentiality of employees to the extent possible while the information is being used for occupational safety and health purposes.

U.S. Department of Labor
Occupational Safety and Health Administration

Year 20____

Form approved OMB no. 1218-0176

You must record information about every work-related death and about every work-related injury or illness that involves loss of consciousness, restricted work activity or job transfer, days away from work, or medical treatment beyond first aid. You must also record significant work-related injuries and illnesses that are diagnosed by a physician or licensed health care professional. You must also record work-related injuries and illnesses that meet any of the specific recording criteria listed in 29 CFR Part 1904.8 through 1904.12. Feel free to use two lines for a single case if you need to. You must complete an Injury and Illness Incident Report (OSHA Form 301) or equivalent form for each injury or illness recorded on this form. If you're not sure whether a case is recordable, call your local OSHA office for help.

Establishment name _____

City _____ State _____

Identify the person

(A) Case no.	(B) Employee's name	(C) Job title (e.g., Welder)

Describe the case

(D) Date of injury or onset of illness	(E) Where the event occurred (e.g., Loading dock north end)	(F) Describe injury or illness, parts of body affected, and object/substance that directly injured or made person ill (e.g., Second degree burns on right forearm from acetylene torch)
month/day		
month/day		
month/day		
month/day		
month/day		
month/day		
month/day		
month/day		
month/day		
month/day		
month/day		
month/day		
month/day		

Classify the case

Using these four categories, check ONLY the most serious result for each case:

Death (G)	Days away from work (H)	Remained at work: Job transfer or restriction (I)	Other recordable cases (J)

Enter the number of days the injured or ill worker was:

On job transfer or restriction (K)	Away from work (L)
____ days	____ days
____ days	____ days
____ days	____ days
____ days	____ days
____ days	____ days
____ days	____ days
____ days	____ days
____ days	____ days
____ days	____ days
____ days	____ days
____ days	____ days
____ days	____ days
____ days	____ days

Check the "Injury" column or choose one type of illness:

(M)

Injury (1)	Skin disorder (2)	Respiratory condition (3)	Poisoning (4)	All other illnesses (5)

Page totals ▶

Be sure to transfer these totals to the Summary page (Form 300A) before you post it.

	Injury (1)	Skin disorder (2)	Respiratory condition (3)	Poisoning (4)	All other illnesses (5)

Page ____ of ____

Public reporting burden for this collection of information is estimated to average 14 minutes per response, including time to review the instructions, search and gather the data needed, and complete and review the collection of information. Persons are not required to respond to the collection of information unless it displays a currently valid OMB control number. If you have any comments about these estimates or any other aspects of this data collection, contact: US Department of Labor, OSHA Office of Statistics, Room N-3644, 200 Constitution Avenue, NW, Washington, DC 20210. Do not send the completed forms to this office.

OSHA's Form 300A

Summary of Work-Related Injuries and Illnesses

Year 20__

U.S. Department of Labor
Occupational Safety and Health Administration

Form approved OMB no. 1218-0176

All establishments covered by Part 1904 must complete this Summary page, even if no work-related injuries or illnesses occurred during the year. Remember to review the Log to verify that the entries are complete and accurate before completing this summary.

Using the Log, count the individual entries you made for each category. Then write the totals below, making sure you've added the entries from every page of the Log. If you had no cases, write "0."

Employees, former employees, and their representatives have the right to review the OSHA Form 300 in its entirety. They also have limited access to the OSHA Form 301 or its equivalent. See 29 CFR Part 1904.35, in OSHA's recordkeeping rule, for further details on the access provisions for these forms.

Number of Cases

Total number of deaths	Total number of cases with days away from work	Total number of cases with job transfer or restriction	Total number of other recordable cases
____ (G)	____ (H)	____ (I)	____ (J)

Number of Days

Total number of days of job transfer or restriction	Total number of days away from work
____ (K)	____ (L)

Injury and Illness Types

Total number of . . .
(M)

(1) Injuries ____
(2) Skin disorders ____
(3) Respiratory conditions ____

(4) Poisonings ____
(5) All other illnesses ____

Establishment information

Your establishment name _____

Street _____

City _____ State _____ ZIP _____

Industry description (e.g., Manufacture of motor truck trailers)
_____

Standard Industrial Classification (SIC), if known (e.g., SIC 3715)
__ __ __ __

Employment information (If you don't have these figures, see the Worksheet on the back of this page to estimate.)

Annual average number of employees _____

Total hours worked by all employees last year _____

Sign here

Knowingly falsifying this document may result in a fine.

I certify that I have examined this document and that to the best of my knowledge the entries are true, accurate, and complete.

_____   _____
Company executive   Title

(____) ____ - ____   ____ / ____ / ____
Phone   Date

Post this Summary page from February 1 to April 30 of the year following the year covered by the form.

Public reporting burden for this collection of information is estimated to average 50 minutes per response, including time to review the instructions, search and gather the data needed, and complete and review the collection of information. Persons are not required to respond to the collection of information unless it displays a currently valid OMB control number. If you have any comments about these estimates or any other aspects of this data collection, contact: US Department of Labor, OSHA Office of Statistics, Room N-3644, 200 Constitution Avenue, NW, Washington, DC 20210. Do not send the completed forms to this office.

Optional

Worksheet to Help You Fill Out the Summary

At the end of the year, OSHA requires you to enter the average number of employees and the total hours worked by your employees on the summary. If you don't have these figures, you can use the information on this page to estimate the numbers you will need to enter on the Summary page at the end of the year.

How to figure the average number of employees who worked for your establishment during the year:

❶ Add the total number of employees your establishment paid in all pay periods during the year. Include all employees: full-time, part-time, temporary, seasonal, salaried, and hourly.

The number of employees
paid in all pay periods =   ❶ _____

❷ Count the number of pay periods your establishment had during the year. Be sure to include any pay periods when you had no employees.

The number of pay
periods during the year =   ❷ _____

❸ Divide the number of employees by the number of pay periods.

$$\frac{❶}{❷} = ❸ \_\_\_\_\_$$

❹ Round the answer to the next highest whole number. Write the rounded number in the blank marked *Annual average number of employees.*

The number rounded = ❹ _____

For example, Acme Construction figured its average employment this way:

For pay period…	Acme paid this number of employees.
1	10
2	0
3	15
4	30
5 ▶	40
24	20
25	15
26	+10
	830

Number of employees paid = 830   ❶

Number of pay periods = 26   ❷

$\dfrac{830}{26}$ = 31.92   ❸

31.92 rounds to 32   ❹

32 is the annual average number of employees

How to figure the total hours worked by all employees:

Include hours worked by salaried, hourly, part-time and seasonal workers, as well as hours worked by other workers subject to day to day supervision by your establishment (e.g., temporary help services workers).

Do not include vacation, sick leave, holidays, or any other non-work time, even if employees were paid for it. If your establishment keeps records of only the hours paid or if you have employees who are not paid by the hour, please estimate the hours that the employees actually worked.

If this number isn't available, you can use this optional worksheet to estimate it.

Optional Worksheet

Find the number of full-time employees in your establishment for the year.

_____

Multiply by the number of work hours for a full-time employee in a year.

X _____

This is the number of full-time hours worked.

_____

Add the number of any overtime hours as well as the hours worked by other employees (part-time, temporary, seasonal)

+ _____

Round the answer to the next highest whole number. Write the rounded number in the blank marked *Total hours worked by all employees last year.*

_____

OSHA's Form 301
Injury and Illness Incident Report

U.S. Department of Labor
Occupational Safety and Health Administration

Form approved OMB no. 1218-0176

Attention: This form contains information relating to employee health and must be used in a manner that protects the confidentiality of employees to the extent possible while the information is being used for occupational safety and health purposes.

This *Injury and Illness Incident Report* is one of the first forms you must fill out when a recordable work-related injury or illness has occurred. Together with the *Log of Work-Related Injuries and Illnesses* and the accompanying *Summary*, these forms help the employer and OSHA develop a picture of the extent and severity of work-related incidents.

Within 7 calendar days after you receive information that a recordable work-related injury or illness has occurred, you must fill out this form or an equivalent. Some state workers' compensation, insurance, or other reports may be acceptable substitutes. To be considered an equivalent form, any substitute must contain all the information asked for on this form.

According to Public Law 91-596 and 29 CFR 1904, OSHA's recordkeeping rule, you must keep this form on file for 5 years following the year to which it pertains.

If you need additional copies of this form, you may photocopy and use as many as you need.

Completed by _____

Title _____

Phone (_____) _____ - _____ Date ____ / ____ / ____

Information about the employee

1) Full name _____

2) Street _____
   City _____ State _____ ZIP _____

3) Date of birth ____ / ____ / ____

4) Date hired ____ / ____ / ____

5) ☐ Male
   ☐ Female

Information about the physician or other health care professional

6) Name of physician or other health care professional _____

7) If treatment was given away from the worksite, where was it given?
   Facility _____
   Street _____
   City _____ State _____ ZIP _____

8) Was employee treated in an emergency room?
   ☐ Yes
   ☐ No

9) Was employee hospitalized overnight as an in-patient?
   ☐ Yes
   ☐ No

Information about the case

10) Case number from the Log _____ (Transfer the case number from the Log after you record the case.)

11) Date of injury or illness ____ / ____ / ____

12) Time employee began work ____ : ____ AM / PM

13) Time of event ____ : ____ AM / PM ☐ Check if time cannot be determined

14) **What was the employee doing just before the incident occurred?** Describe the activity, as well as the tools, equipment, or material the employee was using. Be specific. *Examples:* "climbing a ladder while carrying roofing materials"; "spraying chlorine from hand sprayer"; "daily computer key-entry."

15) **What happened?** Tell us how the injury occurred. *Examples:* "When ladder slipped on wet floor, worker fell 20 feet"; "Worker was sprayed with chlorine when gasket broke during replacement"; "Worker developed soreness in wrist over time."

16) **What was the injury or illness?** Tell us the part of the body that was affected and how it was affected; be more specific than "hurt," "pain," or sore." *Examples:* "strained back"; "chemical burn, hand"; "carpal tunnel syndrome."

17) **What object or substance directly harmed the employee?** *Examples:* "concrete floor"; "chlorine"; "radial arm saw." If this question does not apply to the incident, leave it blank.

18) **If the employee died, when did death occur?** Date of death ____ / ____ / ____

Public reporting burden for this collection of information is estimated to average 22 minutes per response, including time for reviewing instructions, searching existing data sources, gathering and maintaining the data needed, and completing and reviewing the collection of information. Persons are not required to respond to the collection of information unless it displays a current valid OMB control number. If you have any comments about this estimate or any other aspects of this data collection, including suggestions for reducing this burden, contact: US Department of Labor, OSHA Office of Statistics, Room N-3644, 200 Constitution Avenue, NW, Washington, DC 20210. Do not send the completed forms to this office.

If You Need Help...

If you need help deciding whether a case is recordable, or if you have questions about the information in this package, feel free to contact us. We'll gladly answer any questions you have.

▼ **Visit us online at www.osha.gov**

▼ **Call your OSHA Regional office and ask for the recordkeeping coordinator**

or

▼ **Call your State Plan office**

Federal Jurisdiction

Region 1 - 617 / 565-9860
Connecticut; Massachusetts; Maine; New Hampshire; Rhode Island

Region 2 - 212 / 337-2378
New York; New Jersey

Region 3 - 215 / 861-4900
DC; Delaware; Pennsylvania; West Virginia

Region 4 - 404 / 562-2300
Alabama; Florida; Georgia; Mississippi

Region 5 - 312 / 353-2220
Illinois; Ohio; Wisconsin

Region 6 - 214 / 767-4731
Arkansas; Louisiana; Oklahoma; Texas

Region 7 - 816 / 426-5861
Kansas; Missouri; Nebraska

Region 8 - 303 / 844-1600
Colorado; Montana; North Dakota; South Dakota

Region 9 - 415 / 975-4310

Region 10 - 206 / 553-5930
Idaho

State Plan States

Alaska - 907 / 269-4957

Arizona - 602 / 542-5795

California - 415 / 703-5100

*Connecticut - 860 / 566-4380

Hawaii - 808 / 586-9100

Indiana - 317 / 232-2688

Iowa - 515 / 281-3661

Kentucky - 502 / 564-3070

Maryland - 410 / 767-2371

Michigan - 517 / 322-1848

Minnesota - 651 / 284-5050

Nevada - 702 / 486-9020

*New Jersey - 609 / 984-1389

New Mexico - 505 / 827-4230

*New York - 518 / 457-2574

North Carolina - 919 / 807-2875

Oregon - 503 / 378-3272

Puerto Rico - 787 / 754-2172

South Carolina - 803 / 734-9669

Tennessee - 615 / 741-2793

Utah - 801 / 530-6901

Vermont - 802 / 828-2765

Virginia - 804 / 786-6613

Virgin Islands - 340 / 772-1315

Washington - 360 / 902-5554

Wyoming - 307 / 777-7786

*Public Sector only

U.S. Department of Labor
Occupational Safety and Health

Have questions?

If you need help in filling out the *Log* or *Summary*, or if you have questions about whether a case is recordable, contact us. We'll be happy to help you. You can:

▶ Visit us online at: **www.osha.gov**

▶ Call your regional or state plan office. You'll find the phone number listed inside this cover.

U.S. Department of Labor
Occupational Safety and Health Administration

Sample Jobsite Safety Audit Checklist

	Yes	No	N/A

Administration

	Yes	No	N/A
OSHA Form 300 (if required)	☐	☐	☐
OSHA safety and health poster	☐	☐	☐
Chemical information list	☐	☐	☐
Material Safety Data Sheets	☐	☐	☐
Trade contractors' chemical information lists	☐	☐	☐
Emergency phone numbers	☐	☐	☐
Worker's compensation poster	☐	☐	☐

Personal Protective Equipment Issued/Used as Instructed

	Yes	No	N/A
Hard hats	☐	☐	☐
Protective glasses/goggles/face shields	☐	☐	☐
Gloves	☐	☐	☐
Respirators	☐	☐	☐
Hearing protection	☐	☐	☐
Work shoes/boots	☐	☐	☐

Ladders

	Yes	No	N/A
In good condition	☐	☐	☐
Properly extended and secured	☐	☐	☐
Extended 36 inches above roof or platform (when used for access)	☐	☐	☐
Doors blocked open, locked, or guarded when in front of ladder	☐	☐	☐
Stepladders fully open when used	☐	☐	☐
Damaged ladders removed from service	☐	☐	☐
Proper ladder used for the task	☐	☐	☐
Used where there is a break of more than 19 inches	☐	☐	☐

Housekeeping and Sanitation

	Yes	No	N/A
Adequate trash removal	☐	☐	☐
Floor openings covered or guarded	☐	☐	☐
Stairs and walkways clear of debris and materials	☐	☐	☐

	Yes	No	N/A
Guardrails erected on stairways, wall openings, etc.	☐	☐	☐
Adequate lighting provided	☐	☐	☐
Toilet facilities adequate	☐	☐	☐
Washing facility available	☐	☐	☐
Drinking water and cups provided	☐	☐	☐
Phone available for emergency calls	☐	☐	☐

Scaffolding

	Yes	No	N/A
Sound footing for all platforms	☐	☐	☐
Proper platform used	☐	☐	☐
Safe access to all working levels	☐	☐	☐
Equipped with standard guardrails, midrails, and toeboards	☐	☐	☐
Head protection provided where workers work or pass under scaffolding in use	☐	☐	☐
No accumulation of tools or material on platforms	☐	☐	☐
Scaffold-specific rules followed	☐	☐	☐

Power and Hand Tools

	Yes	No	N/A
Proper tool used for job performed	☐	☐	☐
Guard and safety devices are operable and in place	☐	☐	☐
Tool retainers used on pneumatic tools; air pressure properly regulated	☐	☐	☐
Pinch and shear points guarded	☐	☐	☐
Employees using/operating tools safely	☐	☐	☐
Personal protective equipment in use	☐	☐	☐
Safety device installed on hoses at the source of supply or branch line (for hoses exceeding 3½ inches)	☐	☐	☐

Powder-Actuated Tools

	Yes	No	N/A
All operators trained and certified	☐	☐	☐
Tools and charges protected from unauthorized use	☐	☐	☐

	Yes	No	N/A

Power-Actuated Tools (*Continued*)

	Yes	No	N/A
Loaded tools not left unattended	☐	☐	☐
All tools inspected and tested daily before use	☐	☐	☐
Tools and charges matched to recommended materials only	☐	☐	☐
Safety eyewear or face shields used by operators	☐	☐	☐

Motor Vehicles

	Yes	No	N/A
Audible back-up alarms	☐	☐	☐
Horns maintained	☐	☐	☐
Brake systems maintained	☐	☐	☐
Regular inspections conducted	☐	☐	☐
Roll-over protection structure maintained	☐	☐	☐

Welding and Cutting

	Yes	No	N/A
All operators trained and qualified	☐	☐	☐
Personal protective equipment used	☐	☐	☐
Fire extinguishers provided	☐	☐	☐
Cylinders secured	☐	☐	☐
All fittings free of oil and grease	☐	☐	☐
Proper gauge settings maintained	☐	☐	☐
All hoses, cords, and other equipment in good condition	☐	☐	☐

Material Storage and Handling

	Yes	No	N/A
Materials properly stacked and on firm footing	☐	☐	☐
Fire protection adequate	☐	☐	☐
Employees handling loads properly	☐	☐	☐
Flammable liquids stored only in approved containers	☐	☐	☐

Fall Protection

	Yes	No	N/A
Guardrails and toeboards installed where necessary	☐	☐	☐
Personal fall-arrest systems used where necessary	☐	☐	☐
Alternative safe work practices with a controlled access zone when required	☐	☐	☐

	Yes	No	N/A
Floor holes and skylights covered or guarded	☐	☐	☐

Excavation and Trenching

	Yes	No	N/A
All excavation trenches properly shored or sloped according to a competent person's recommendations	☐	☐	☐
Excavations have proper means of entry/exit	☐	☐	☐
Excavations/trenches inspected by a competent person	☐	☐	☐
Water accumulation avoided	☐	☐	☐
Basement foundation trenches adequately protected	☐	☐	☐

Electricity

	Yes	No	N/A
Ground fault circuit interrupters used with temporary power	☐	☐	☐
Tools are double insulated	☐	☐	☐
Extension cords are junior/hard-service type	☐	☐	☐
At least 10 feet of distance maintained from all overhead power lines	☐	☐	☐
Extension cords in safe condition	☐	☐	☐

First-Aid Kits

	Yes	No	N/A
Kits provided	☐	☐	☐
Kits inspected and replenished when necessary	☐	☐	☐
Training on use of first-aid kit items and on first-aid procedures regularly conducted and up to date	☐	☐	☐

Fire Prevention

	Yes	No	N/A
Fire extinguishers provided and suitable for onsite hazards	☐	☐	☐
Portable fire extinguishers operational	☐	☐	☐
No-smoking signs posted	☐	☐	☐

	Yes	No	N/A
Safety cans used for flammable liquids of more than 1 gallon	☐	☐	☐
Stairways and exits clear of flammable or combustible materials	☐	☐	☐
All liquid propane tanks stored outside	☐	☐	☐

Hazardous Materials

	Yes	No	N/A
Properly stored	☐	☐	☐
Properly labeled	☐	☐	☐
Label maintained	☐	☐	☐
Hazcom program followed	☐	☐	☐
Proper personal protective equipment used	☐	☐	☐

Project/site name _____

Project number _____

Superintendent _____

Date _____

Sample Plan for Hazard Communications

1. **The Hazcom standard.** Employees are entitled to be informed about any hazardous materials with which they will come into contact in the workplace. They are entitled to receive information about what the materials are and the hazards they pose. They are entitled to obtain guidance, instruction, and/or appropriate personal protective equipment to help protect against those dangers. They are entitled to see information about the materials.

   This standard is communicated to employees as part of their Employee handbook.

2. **Material Safety Data Sheets.** The suppliers of hazardous materials must supply Material Safety Data Sheets (MSDSs) for the materials. The MSDSs are available for employee inspection and review at any time. [*Note:* If you do not receive a MSDS, you are required to request one.] The MSDSs are kept at this location:

   _____

3. **List of known hazardous materials.** The following hazardous materials may be found in our worksites:

   _____

   _____

4. **Protective equipment.** The following personal protective equipment is available and required when working with any of the hazardous materials noted above:

   _____

   _____

5. **Safety procedures and training.** Employee safety training is covered by the following steps:

   _____

   _____

6. **Policy on hazardous materials brought into workplace by other employers on worksites.** Employees will be notified and trained regarding these hazardous materials by the following steps:

   _____

   _____

Personnel Records Form

Date received

- [ ] Hire date_____
- [ ] Pay change
- [ ] Other

- [ ] Separation date_____
- [ ] Administrative Adjustment

- [ ] Records Change
- [ ] Job change

Employee I.D. Data

Last name _____ First name _____ Social Security Number

Employee number _____ Dept./Location _____

- [ ] Employee I-9 Verification _____

Employee Personal Data

Home Address _____ Date of Birth _____

_____ Zip code _____

Emergency Contact _____ Relationship _____

Address _____

Home phone _____ Work phone _____ E-mail _____

Employment Data

Job Title _____ Supervisor _____

Status: [ ] Full time   [ ] Part time   [ ] Temporary

FLSA Status   [ ] Non-exempt   [ ] Exempt (test filed)

EEO Job class _____

Pay Records

Current pay rate $_____ per [ ] hour [ ] year [ ] other

- [ ] W-4 exemptions change

New pay rate $_____ per [ ] hour [ ] year [ ] other

Effective date _____

Appraisal Review

_____

_____

_____

Benefits

Rating: [ ] Exceeds expectations [ ] Meets expectations

[ ] Does not meet expectations

[ ] Other _____ Next review _____

Leave

- [ ] Vacation   [ ] Disability/Pregnancy   [ ] Family Medical
- [ ] Other _____

Start date _____ Return date _____

- [ ] Approved _____ Date _____

Employee Signature _____ Date _____ Supervisor Signature _____ Date

Other Signature _____ Date

Sample Performance Appraisal Form

Name_____ Position _____

Department_____ Hire Date _____

Appraisal Period From_____ To _____

Supervisor_____

Performance Ratings:

1 Excellent: Employee provides exceptional contribution. Provides best performance possible for the job. Substantially exceeds goals and responsibilities.
2 Very good: Employee meets all expectations and exceeds some goals and responsibilities.
3 Satisfactory: Employee meets all required goals and responsibilities.
4 Improvement needed: Employee's performance is regularly below job standards.
5 Unsatisfactory: Employee does not follow set policies and procedures. Fails to meet required job responsibilities and duties.

	Rating	**Comments**
JOB KNOWLEDGE: Understands job functions and how position relates to others in the department and within the company.		
PRODUCTIVITY: Produces work at a pace and quality level that meets job requirements.		
TIME MANAGEMENT: Has good organizational skills: Meets deadlines, sets priorities, uses time wisely.		
ATTITUDE: Cooperates with supervisor, co-workers, and others; uses tact; is enthusiastic about job.		
JOB IMPROVEMENT: Receptive to new ideas; accepts change easily.		
DEPENDABILITY: Can be depended on to get the job done as requested and in a timely fashion.		

	Rating	Comments
ATTENDANCE: Arrives at work punctually; has good attendance record.		
RESPONSIBILITY: Performs regular job functions and, when workload permits, seeks additional responsibility.		
COMPANY POLICIES: Adheres to policies and procedures (e.g., safety, conduct, etc.).		
JUDGMENT: Recognizes consequences of decisions, demonstrates good judgment in thinking, timing, insight.		
COMMUNICATION: Expresses self logically and clearly; communicates effectively with others.		
SUPERVISION (if applicable): Trains others and monitors workflow; delegates and coordinates effectively.		
OVERALL PERFORMANCE RATING:		

Strengths (cite examples):

_____

Areas needing improvement (cite specific examples for each problem area and actions to be taken):

_____

Training completed in the past year:

Training to be completed in the upcoming year:

_____

Goals to be met in the upcoming year:

_____

Supervisor's comments:

_____

If overall rating is unsatisfactory, has employee been warned that performance must be improved?
☐ Yes   ☐ No

Steps to be taken to improve performance

_____

Employee's comments:

1. Employee's comments about the performance review:

_____

2. Goals: During the next year, I would like to achieve (These goals have been set and agreed upon by the employee and his or her supervisor):

_____

I acknowledge receipt of this evaluation and it has been discussed with me. My signature does not imply that I agree or disagree with the content. I acknowledge that nothing within this review modifies my at-will employment status.

_____   _____
Employee Signature   Date   Supervisor Signature   Date

_____
President Signature   Date

Sample Performance Review Preparation Form

TO: [Employee's name]

FROM: _____

RE: Review of Employee Performance

Date and Time of Performance Review _____

You are not required to complete this form and it is not to be turned in to your supervisor. However, it will help you prepare for you performance review.

_____

_____

What do you feel are your greatest performance strengths?

_____

_____

In what areas do you feel you need to improve?

_____

_____

What questions or comments do you have concerning your current position?

_____

_____

What improvements would you like to see made in the methods and procedures related to performing your job duties?

_____

_____

What goals would you like to achieve in the coming year?

_____

_____

In your opinion, which of your skills could be better utilized in your current position?

_____

_____

What training would help you perform your current job duties?

_____

_____

Please use another page for additional comments, if necessary, and bring this form to your performance review discussion.

Sample Notice of Unsatisfactory Performance

Employee Name _____

Department/Location _____ Date _____

This notice advises you that you are not meeting minimal acceptable standards of performance for your job. Continued below-standard performance may result in disciplinary action and/or termination. Immediate improvement in performance is needed.

Performance standard(s) not being met _____

_____

Specific example(s) of your unsatisfactory performance _____

_____

Improvements needed _____

_____

Prior notice of problem to employee?   ☐ Verbal   ☐ Written

Improvement review date _____

If improvement is not made by this date, the employee is subject to:

☐ Second written warning   ☐ Suspension without pay   ☐ Termination

Employee comments _____

_____

_____

_____

_____   _____
Employee Signature   Date   Supervisor Signature   Date

Sample Termination Checklist

Employee name _____

Date of termination _____

The following items can be covered as part of an employee termination procedure. Check off each item as it is completed. Retain this checklist in the employee's personnel folder.

☐ 1. Exit interview   ☐ Personal interview conducted (Date)
(Check one)

☐ Interview form given to employee to return by mail

☐ Involuntary separation

☐ 2. Company property collected (check as many as apply)

☐ Keys   ☐ Company reference materials

☐ Tools and equipment   ☐ Company information

☐ Employee handbook   ☐ Employee identification cards

☐ Other _____

☐ Company property still outstanding _____

_____

☐ 3. Final payroll and accounting

☐ Any advances or loans to be collected?

☐ Final paycheck procedure discussed; to be issued

☐ Vacation hours payment discussed

☐ 4. Continuation of medical insurance notice given

☐ Completed form returned   Date _____

_____   _____
Completed by   Date

_____   _____
Employee Signature   Date

Sample Exit Interview Form

Employee _____ Date _____

Type of termination: ☐ Resignation   ☐ Firing   ☐ Layoff   ☐ Other

Stated reason for resignation: _____

☐ Obtained another position with _____

Position _____ Compensation $ _____

☐ Returning to school
☐ Personal _____
☐ Wanted to leave the company
☐ Other _____

What did you like best about your job?

_____

_____

What did you like least about your job?

_____

_____

What did you like best about working here?

_____

_____

What did you like least about working at here?

_____

_____

Any final comments or suggestions that could help this company be a better employer?

_____

_____

_____
Employee Signature   Date